Contents

iii

Acknowledgements

The completion of this Handbook has been possible through the generous support and encouragement of the staff and students of the Department of Earth Sciences at Goldsmith's College. In particular I thank Dr D. Helm for his critical reviews and advice. T. Easter ably assisted with the photographic reproductions, M. Insley assisted with the drafting of the figures and Carol Sutton assisted with the word processing. Dr M. de Freitas, the first *Geological Society of London Handbook Series* Editor and Ms R. Dixon of the Open University Press are thanked for their advice, patience and encouragement.

A first draft of this handbook was prepared whilst the author was on sabbatical leave at the Geological Survey of Canada, Vancouver, supported by the Royal Society and the Geological Association of Canada. Dr R. I. Thompson is thanked for his support and encouragement during this period. M. de Freitas, P. Ransom, P. Ellis and M. Insley gave critical reviews.

Illustrations in this Handbook were adapted from the following sources to whom grateful acknowledgement is made:

Fig. 2.2: RAGAN D. M. (1985): *Structural Geology; an Introduction to Geometric Techniques.* 3rd ed., John Wiley and Sons, New York.
Fig. 3.5: HUDDLESTON, P., (1973): *Tectonophysics* 16, 1–46. Amsterdam, Elsevier.
Fig. 3.6 & Fig. 3.7: WILLIAMS G. D. & CHAPMAN, T. J. (1979), *Journal of Structural Geology,* 1, 181–186, Oxford, Pergamon.
Fig. 3.9 & Fig. 3.10: BELL, A. M., (1981), *Journal of Structural Geology,* 3, 197–202. Oxford, Pergamon.
Fig. 6.23b: RAMSAY, J. G., (1980). *Journal of Structural Geology,* 2, 83–99, Oxford, Pergamon.
Table 6.4: SIBSON, R. H., (1977). *Journal of the Geological Society of London,* 133, 191–214. Oxford, Blackwells.
Fig. 9.1: RAGAN, D. M., (1985). *Structural Geology; an Introduction to Geometric Techniques.* 3rd ed., New York, John Wiley and Sons.
Photographs for Figs. 4.1d and 7.7 were kindly supplied by John Ramsay.

1

Introduction

1.1 Objectives

This book is designed as a basic guide to the field mapping and interpretation of geological structures. Emphasis is placed upon the identification of structures and the systematic recording of structural data, as both should be a fundamental part of any mapping programme. The identification and description of structures, together with an understanding of their development, i.e. their movement patterns (*Kinematic analysis*) and an appreciation of the forces and stresses responsible for them (*Dynamic analysis*) are extremely useful for interpreting particular structures, and for knowing what geometry to expect whilst mapping in a particular terrane.

Structural data cannot be recorded or used in a vacuum. They must be accompanied by full lithological, sedimentological, petrological and palaeontological descriptions for their complete interpretation.

The following aspects are emphasised in this Handbook;

1 Recognition of structures.
2 What to measure and what to describe.
3 How to analyse the data collected.

4 How to interpret the data and incorporate it into the stratigraphy, interpretation and regional syntheses for an area.

In all cases emphasis is placed upon systematic field observations, accurate measurements of the orientations of structural elements, careful recording of the data in the field notebook, sketching and photographing the structures, and analysis in the field using the stereographic projection. Above all, structural geology requires the appreciation of the three-dimensional nature of structures. Think in 3D and learn to extend your view of structures above and below the map sheet.

1.2 Fieldwork

The importance of careful, accurate and systematic fieldwork cannot be overemphasised. Basic geologic mapping techniques are described in Barnes (1981), and the field descriptions of sedimentary, metamorphic and igneous rocks are outlined in the companion Handbooks by Tucker (1982), Fry (1984), and Thorpe and Brown (1985) respectively.

This Handbook describes the field techniques for mapping geological structures and for the identification and mapping of particular types of structure. It also gives a brief summary of the interpretation and analysis of structures.

Remember the following points:

1 Accurate measurement, observation and recording of all structural elements is essential. Avoid data selection in the field, otherwise you may find that upon further interpretation in the laboratory, you have failed to measure a key structural feature.

2 Carry out an ongoing interpretation whilst in the field (draw sketch cross-sections and maps). This will help you recognise key areas where further work may be necessary. Your interpretation will be governed by your experience and knowledge of regional structure, *but only accurate and well-recorded data will have a permanent value and permit continuous reinterpretation.*

3 Data should always be plotted on maps and cross-sections whilst in the field. Only in these circumstances can an effective, ongoing interpretation be achieved.

4 Structural data must be collected in conjunction with other lithological, petrological and palaeontological data.

Conduct and safety in the field

Fieldwork frequently puts geologists in hazardous situations. Structural geologists commonly work in rugged and exposed terrain where 3D exposure is good. Be safety conscious and aware of the possible dangers, particularly from loose rock underfoot, and from rock falls. Barnes (1981) outlines fieldwork safety, and in addition to reading this the reader should also consult the safety checklist on p. 16 of this Handbook before commencing fieldwork. Always carry out fieldwork in compliance with the Geologists' Association Code of Conduct (see Barnes, 1981).

1.3 Tectonic and structural regimes

It is beyond the scope of this Handbook to describe regional structural relationships in detail, but it is useful to identify the dominant features associated with particular tectonic settings, as they provide a useful guide to the structures that may be found whilst mapping (Table 1.1). Characteristic families of structures may be expected to occur in a particular environment, e.g. shallow thrust faults and parallel folding in frontal regions of foreland fold and thrust belts, and this knowledge can greatly aid any ongoing interpretation. Table 1.1 is neither exhaustive nor exclusive in its contents and you should always be prepared for other structures to occur and record all the structural information from the outcrops in your mapping area.

2

Table 1.1 Structures associated with particular tectonic regimes. (Cont'd on p. 4 and p. 5)

INTRA PLATE REGIMES

	Passive continental margins (Atlantic type).	Continental rift zones	Intra-plate strike-slip zones	Intra-plate fold and fault belts.
Major structural elements	Extensional (normal) faulting. Syndepositional tectonics, Salt tectonics.	Extensional (normal) faulting. Strike-slip systems linking extensional faults.	Major fault systems, associated en-echelon folding. Secondary extensional and contractional faulting along curved, overlapping fault systems.	Variable folding & thrusting. Extensional faulting associated with regional uplift.
Metamorphism	None to burial metamorphism. Compaction due to burial.	Hydrothermal systems and volcanic activity, elevated heat flow. Dynamic metamorphism associated with faults, cataclasites — mylonites.	Generally low grade.	Variable — to granulite facies. Development of fault rocks, cataclasites — mylonites along active fault zones.
Examples	Eastern U.S.A. continental margin, West African continental margin.	North Sea Basin. East African Rift System.	Northern Rocky Mountain Trench — Tintina Fault System, Canada.	Basin and Range, U.S.A.

Table 1.1 (cont'd) Structures associated with particular tectonic regimes.

ACTIVE PLATE MARGIN REGIMES

	Constructive	Conservative	Destructive	Collision
Major structural elements	Mid-ocean ridge systems, and marginal basin spreading systems	Major strike-slip fault systems	Island arc or continental margin arc systems	Continent–continent or continent – island arc collision
	Extensional (normal) fault systems, major strike-slip (transform) fault systems	Strike-slip fault systems local extension (normal) and contractional (reverse or thrust) fault systems. Local folding — typically en-echelon patterns. Development of pull-apart basins along fault system.	Subduction complexes — Fold and thrust belts — uplifted volcanic arcs — Fore-arc basins, oblique subduction — strike-slip systems. *Subduction complexes* — Contractional (thrust) faulting, Vein systems, penetrative cleavages, melanges. *Fold and thrust belts* — Thrust and fold nappes — *Uplifted volcanic arcs* — extensional faults, fracture patterns associated with intrusions and volcanics. *Fore-arc basins* — local extensional tectonics.	Major overthrust sheets, (allochthonous). Major fold nappes. Major strike-slip faults. *In internal zones* — Fold nappes, contractional (thrust) faults, polyphase deformation. Major strike-slip faults, uplift and late extensional (normal) faults. *In external zones* — Foreland fold and thrust belts, Minor strike-slip faults (generally simpler geometry than internal zones). Development of foreland basins which may become involved in the thrusting.

4

Table 1.1 (cont'd) Structures associated with particular tectonic regimes.

ACTIVE PLATE MARGIN REGIMES (Cont'd)

	Constructive	Conservative	Destructive	Collision
Metamorphism	Range of Metamorphism from Zeolite, Greenschist, Amphibolite. Hydrothermal alteration and vein systems.	Low-grade — sub greenschist burial metamorphism. Local dynamic metamorphism (cataclasites mylonites) and hydrothermal alteration along major fault zones.	High pressure low temperature metamorphism in subduction complexes. Low pressure high temperature metamorphism in interior of arc (associated with intrusions).	*Internal zones* — high-grade polymetamorphism and igneous intrusions, penetrative foliations. *External zones* — Low-grade or burial metamorphism, one or no penetrative foliation.
Examples	Icelandic Rift System	San Andreas Fault System, Dead Sea Transform System	Japanese Island Arc Systems	Himalayan Collision Zone

5

1.4 Bedding

In sedimentary and many metamorphic rocks, *bedding surfaces* (surfaces of primary accumulation) are our *principal reference frame* (or datum). There are many possible bedforms in sedimentary sequences (see Tucker, 1982 for details) and the structural geologist must be aware that significant departures from layer-parallel stratigraphy can occur in certain sedimentary environments, e.g. deltas; thus structural data must always be collected in conjunction with sedimentological and stratigraphic data.

Bedding is one of the most important structural elements and the structural data that should be collected for bedding are outlined in Table 1.2. *The spatial distribution of bedding or compositional banding (e.g. in gneissic terranes), will define the major fold and fault structures within your mapping area.*

1.4.1 Way-up/younging and facing

Way-up/younging is the direction in which stratigraphically younger beds/units are found. (The term *tops* is also sometimes used in this context.)

The *stratigraphic way-up* is of fundamental importance in determining the structure of an area. It is based upon a knowledge of stratigraphy and of small-scale sedimentary structures which indicate the stratigraphic way-up and the sequence of deposition. Sedimentary structures which indicate way-up are discussed in Tucker, 1982 and are summarised in Fig. 1.1. Always look for and record way-up features when mapping.

The *structural way-up* refers to the bedding/cleavage relationships that indicate the position within a major fold structure (e.g. on the overturned limb of a recumbent fold). This may have no relationship to stratigraphic way-up. Take care to distinguish the two — see Chapter 3 for greater detail.

Facing is the direction within a structure i.e. along the fold axial plane or cleavage plane, in which younger beds/units are found. This term is generally applied to folds, or cleavage relationships.

1.5 'Synsedimentary' versus tectonic structures

In many areas of deformed sedimentary rocks it is difficult to distinguish between structures formed during deposition or early diagenesis when the sediment was unconsolidated, and those formed after lithification in response to tectonic forces. On cursory examination many 'synsedimentary' structures such as slump folds have superficial geometric similarities to 'tectonic' folds (Fig. 1.2a). Syndepositional faults are also common (Fig. 1.2b) and in some instances syndepositional cleavage fabrics have been observed (Fig. 1.2c). It is therefore extremely important when mapping to distinguish between syndepositional (prelithification)

Table 1.2 Data to be collected from observations on bedding S_0.

Structure	What to Measure	What Observations to Record	Results of Analysis
Bedding 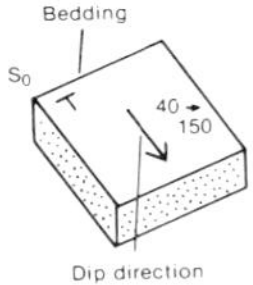	Dip direction (or strike and dip) (Figs. 2.5,2.6, 2.7).	Lithology, bedding thicknesses Grain-size. Grain shapes, grain fabrics.	
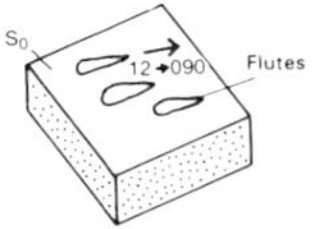	Orientation of sedimentary structures (Figs. 2.11–2.13).	Sedimentary structures. Geopetal structures.	Depositional surfaces. Palaeocurrent directions. Palaeoenvironments. Way-up — younging.
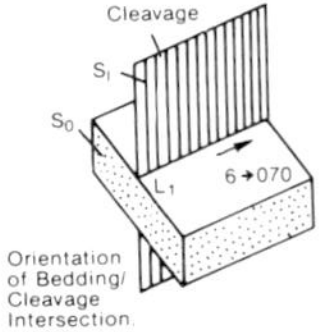	Orientation of tectonic structures on bedding plane (particularly the bedding cleavage intersection) (Figs. 2.11–2.13).	Tectonic structures (cleavage relationships, lineations on bedding plane). (Fig. 4.3b)	Orientation of tectonic structures relative to bedding. (Figs. 4.3b, 5.1b)
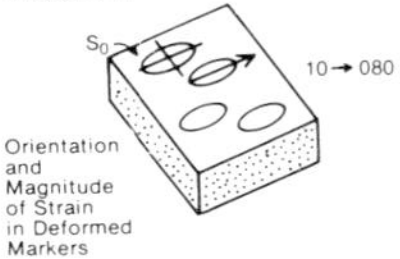	Orientation and magnitude of strain in deformed objects on the bedding plane (Figs. 2.11–2.13 & Appendix III).	Nature of strain relative to bedding. (Fig. 3.12 & Appendix III)	Strain on bedding plane component of layer-parallel shortening; (Fig. 3.12). Relative competencies of units.

Bedding

Dip direction

Orientation of Bedding/ Cleavage Intersection

Orientation and Magnitude of Strain in Deformed Markers

7

DESCRIPTION	PRIMARY STRUCTURE
CROSS-STRATIFICATION Tabular cross-stratification Trough cross–stratification	
NORMAL GRADED BEDDING Coarse grains at the base passing upwards into finer grain sizes; typical of turbidite sequences.	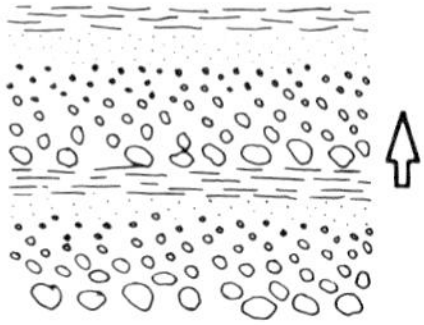
SCOUR STRUCTURES Scour surface at base of sandstone bed overlying mudrock. Coarse-grained lag deposit may occur in the scour.	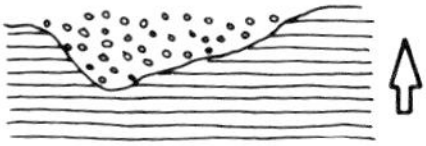
LOAD STRUCTURES Sandstone overlying mudrock Load Casts Flame Stuctures Upward injection of mud into the sandstone	
FLUTE CASTS Developed on the underside of Bedding units in Sandstones. Good Palaeocurrent indicators.	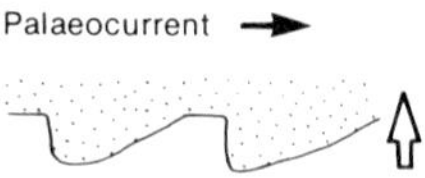 Palaeocurrent ➝

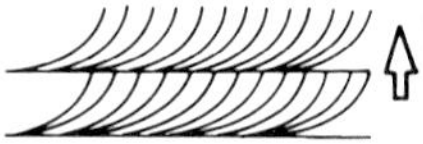

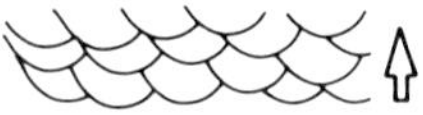

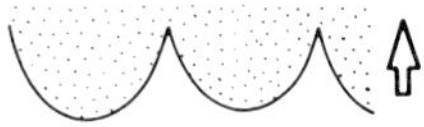

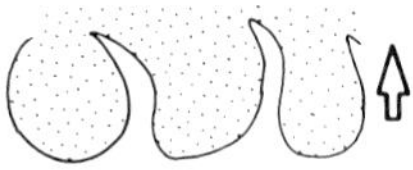

Fig. 1.1 Primary structures that may be used to determine the stratigraphic way-up of beds. (Cont'd on p. 9)

DESCRIPTION	PRIMARY STRUCTURE
Dewatering Structures Pillar structures formed in sandstones and siltstones as water escaped upwards.	
Dewatering Structures Sand volcanoes in mudrocks or siltstones. May be underlain by sandstone dykes.	
Shrinkage Structures Mudcracks infilled with overlying sandstone. Dish structures in mudrock that has undergone desiccation.	
Volcanic Structures Lobate Pillow structures in lavas.	Pillows
Volcanic Structures Blocky, rubbly and weathered tops to lava flows.	blocky top

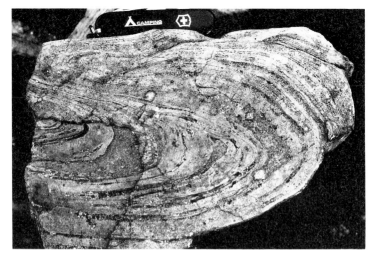

Fig. 1.2a Synsedimentary (Slump) fold in sandstone. Note the absence of fractures, veins, and cleavage.

Fig. 1.2b Listric syndepositional extensional fault (a) in siltstones and mudstones. Note the development of left-dipping antithetic faults (b and c). Scale, near (b), is 1 m.

10

Fig. 1.2c Recumbent Synsedimentary (Slump) fold in siltstones with the development of a weak flat-lying cleavage (cl).

Fig. 1.2d Syndepositional extensional faults (f) in sandstones. Note that the faults are curved in plan and are consistently downthrown to the right. Field of view is 4 m.

11

structures and post-lithification, 'tectonic' structures. In some situations e.g. active continental margins, sediments are deformed by tectonic forces very soon after deposition, before complete lithification. Hence you may find a complete spectrum of structures, from those formed during deposition to those formed deeper in the crust.

1.5.1 Discrimination of prelithification ('synsedimentary') from post-lithification (tectonic) folds

'Synsedimentary' folds or 'slump' folds have many geometric similarities to the shapes, wavelengths and size distributions of 'tectonic folds'. Slump folds are generally tight to isoclinal with variable shapes at low fold amplitudes. Their fold axes are commonly dispersed in the plane of the slump sheet and recumbent folds are dominant (Fig. 1.2c). The fold axial surfaces may be slightly imbricated (stacked up like shingles) and the folds face and verge down the inferred palaeoslope. Axial-planar cleavages are sometimes developed, particularly in the hinge regions (probably due to later compaction during burial). Lineations and grooving are sometimes produced by the motion of the slump sheet, and these may be refolded along with other minor structures. Slump sheet contacts may be gradational. Their upper boundaries may exhibit sharp erosional truncations. Synsedimentary fractures within slumped sequences are gen-

erally not sharp, and fracture openings are not maintained. Veining is absent although the fracture plane may be infilled with mobilised sediment slurry. In general, slump folds have no genetic or geometric relationship to large macroscopic folds.

Syndepositional folding is commonly associated with disturbed sedimentary sequences—synsedimentary extensional faulting, convolute laminations, ball and pillow structures, dewatering structures, sand and mud volcanoes. Remember that slumps are characterised by extensional structures at the rear, whereas the front of the slump sheet will be marked by localised compression, with the development of folds, thrust faults and imbrications. The characteristic features of syndepositional folds are listed in Table 1.3 and are compared with the features of 'tectonic' folds (see Chapter 3).

1.5.2 Discrimination of synsedimentary faulting from tectonic faulting

The attributes of tectonic faults are described in Chapter 6. They are characterised in particular by their geometric relationships with associated structures, folds, fractures and veining, and most importantly, by the development of fault rocks along the fault planes (Chapter 6.6). The presence of faults bounding major basin margins will be revealed by regional mapping, by the associated

Table 1.3 Criteria used to distinguish between synsedimentary folds and 'tectonic folds'. **A** is the most reliable criterion and **C** is the least reliable. Note that several criteria must be used in conjunction in order to determine the origin of a particular fold.

Sedimentary folds	Reliability index	Tectonic folds	Reliability index
Truncation by overlying beds	A	Limited spatial distribution — correlated with regional structure	A
Burrowing or boring by organisms	A		
Cut by synsedimentary dewatering structures	A	Fold vergences and axial planes symmetrical around large folds	A
Undeformed clasts or fossils.	A	Symmetrical fracture patterns and veins — saddle reefs developed	A
Folds with no axial planar cleavage — cut by later tectonic cleavage	B–C	Crystallographic fabrics in non-phyllosilicates (possibly associated with fanning axial-planar cleavage in phyllosilicates)	A
Fold axes strongly dispersed in plane of sheet	B–C		
Dominantly recumbent fold axial planes, may be imbricated (with respect to sheet dip)	C	Slickensides and metamorphic lineations — down fold limbs and around fold hinges	B
Both extensional and compressional features with no veins developed	C	Associated with brittle thrust faults — folds generated by ramps	B
		Kink-like folds with upright axial planes (with respect to the sheet dip)	B
		Continuity of axial planes across several beds	B
		Parasitic relationships between major and minor folds	B
		Fanning axial plane cleavage in phyllosilicates	C

facies distributions of coarse-grained fault-derived sediments adjacent to the fault scarp, by increased sediment thicknesses adjacent to the fault, and by associated smaller syndepositional faults and slumps indicating active tectonism during sedimentation. Here, attention is focussed on outcrop scale features indicative of synsedimentary faults.

1 Synsedimentary faults typically do not affect *all* of the stratigraphic sequence and are overlain by unfaulted sediments in depositional contact.
2 The faults are typically listric in shape (Fig. 1.2b).
3 The faults are typically irregular in plan — often curved (Fig. 1.2d).

4 The down-thrown side of the fault is commonly infilled with a triangular wedge of sediments (Fig. 1.3a) which in some cases may be coarser grained than the surrounding sediments.
5 There is an absence of veining and fault rocks typical of brittle deformation (Chapter 6.6).
6 The fault planes are not generally smooth planar fractures, but are often irregular on a small scale (Fig. 1.3b), commonly with injected sediment slurry along the fault plane.
7 The faults are often associated with syndepositional slumping and disturbed sedimentary sequences — convolute laminations, dewatering structures and sand volcanoes.

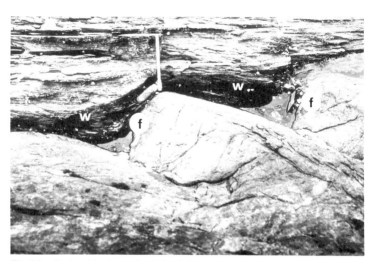

Fig. 1.3a Syndepositional extension faults (f) showing rotation of fault blocks and infilling of the half-graben with a wedge of sediment (w).

14

Fig. 1.3b Small syndepositional fault in sandstones, showing an irregular fault plane and injection of sediment slurry along the fault plane.

1.5.3   Syndepositional cleavage

This is commonly found in the deformed mudstones of slumped sequences. The foliations are planar and parallel to the sheet dip of the sediments, and typically have the appearance of a slaty cleavage or very closely spaced, fine fracture cleavage (Fig. 1.4). They are axial-planar to recumbent slump folds (Fig. 1.2c) and generally do not penetrate sandstone layers but are restricted to mudstones. Slight fanning of the cleavage may occur but the strong refraction of cleavages commonly found in lithified rocks (Chapter 4.3) does not usually occur.

If there is evidence of syn-sedimentary deformation, then great care has to be taken in recognising and mapping cleavage features. In such circumstances careful examination of all of the field relationships is required before a cleavage can be ascribed to syndepositional processes or to later tectonism.

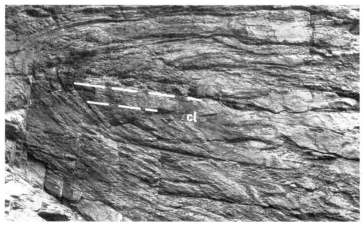

Fig. 1.4   Slump fold with a flat-lying syndepositional axial-planar cleavage (cl). Field of view 2 m.

1.6 Basic References

BADGELY, P. C., (1959) *Structural Methods for the Exploration Geologist*. New York, Harper, 280 pp.

BARNES, J. W., (1981) *Basic Geological Mapping*. Geological Society of London Handbook Series, No. 1. Open University Press, 112 pp.

DAVIS, G. H., (1984) *Structural Geology of Rocks and Regions*. New York, Wiley, 492 pp.

FRY, N., (1984) *The Field Description of Metamorphic Rocks*. Geological Society of London Handbook Series, No. 3. Open University Press, 110 pp.

HOBBS, B. E., MEANS, W. D. & WILLIAMS, P. F., (1976) *An Outline of Structural Geology*. New York, Wiley, 571 pp.

PARK, R. G., (1983) *Foundations of Structural Geology*. Glasgow, Blackie, 135 pp.

PRICE, N. J., (1966) *Fault and Joint Development in Brittle and Semi-Brittle Rock*. Oxford, Pergamon, 176 pp.

RAGAN, D., (1985) *Structural Geology. An Introduction to Geometrical Techniques*. 3rd ed. New York, Wiley, 393 pp.

THORPE, R. S., & BROWN, G. C., (1985) *The Field Description of Igneous Rocks*. Geological Society of London Handbook Series, No. 4. Open University Press, 162 pp.

TUCKER, M., (1982) *The Field Description of Sedimentary Rocks*. Geological Society of London Handbook Series, No. 2. Open University Press, 124 pp.

WILSON, G. (with COSGROVE, J.) (1982) *An Introduction to Small-Scale Geological Structures*. Allen and Unwin, 128 pp.

1.7 Safety

1. Do not run down hills.
2. Do not climb rock faces unless you are a trained climber and you have a friend present.
3. Do not enter old mine workings or cave systems except by arrangement, and always in company.
4. Wear easily seen clothing.
5. Always wear a safety helmet in quarries, under steep cliffs and scree slopes, and underground, and wear goggles when hammering rocks.
6. Note weather forecasts in mountainous country and if you are going into a remote part of an area leave with a responsible person your route map and the time you expect to return.
7. Keep a first aid kit and manual in camp. Carry a small emergency kit in your rucksack, including dressings for blisters, a whistle and a flashlight for signalling (and a mirror if your compass does not have one). Include, also, matches sealed in a waterproof plastic bag, and an aluminized foil 'space blanket' (it weighs almost nothing). In hot climates, carry a water bottle and a packet of effervescent water sterilizing tablets. Always carry some form of emergency ration in case you have to spend a night on a hillside in mist or snow.
8. The accepted field distress signal is six blasts on a whistle or six flashes with a mirror or flashlight, repeated at minute intervals. Rescuers reply with only three blasts or flashes repeated at minute intervals to prevent rescue parties homing in on each other.

2
Mapping techniques

In this section the procedures used in the field to record geological structures will be briefly discussed. Barnes (1981) gives an excellent resume of basic mapping techniques, whereas this Handbook focusses upon structural mapping.

2.1 Equipment

In addition to the usual field equipment—a hammer, hand lens, acid bottle, pen knife and *First Aid kit*, structural mapping requires the following:

In the field

Notebook: Stiff covered and water-proof; sufficiently large to draw sections and sketch maps but not too large to become bulky (20 cms × 10 cms is an optimum size).
Map board: For base maps and/or aerial photographs, non-magnetic (size approximately 30 cm × 25 cm).
Compass-clinometer:Silva Ranger 15 T or a hinged lid type compass with a bubble level e.g. *Freiberg* compass or a *Chaix Universal.* (See p. 18.)
Separate clinometer or level: Optional.

Altimeter: Thommen mountain type for areas with appreciable topography.
Base maps: Detailed topographic maps at the appropriate scales. In structural mapping it is essential to accurately locate yourself and the base maps must have sufficient topographic contours for the purpose—grossly enlarged topographic maps are generally unsuitable. Copies of base maps should be used in the field.
Aerial photographs: are extremely useful even when good base maps are available; in particular for mapping boundary features.
Mylar overlays: for aerial photographs.
Pocket stereoscope.
Good 35 mm camera and films.

In camp

In addition to the necessary drafting equipment, stereographic projections and tracing paper, the following are required:
Graph paper: for section construction.
Map wheel: for section construction.
The appropriate Geological Society Handbooks.

A book on the use of stereographic projections (see References).

Compass-clinometers

For structural mapping you require a compass clinometer that meets the following requirements; (1) accuracy, (2) reliability, (3) ease of operation, and if possible, (4) one that contains a bubble level. Although many types of compass are available (see Barnes, 1981) the following have been found to be the most satisfactory:

(a) the *Silva Ranger* 15 T (Fig. 2.1a)
(b) the *Freiberg* compass (Fig. 2.1b), and
(c) the *Chaix* compass (Fig. 2.1c).

Fig. 2.1a *Silva Ranger* 15TD-CL student's compass/clinometer.

Students will generally use the Silva compass because of its relatively low cost, whereas professional geologists may prefer the Freiberg or Chaix compasses. For structural geology the Freiberg compass has distinct advantages, in that the hinged lid (Fig. 2.1b) is placed against or along-

Fig. 2.1b The Freiberg compass with a hinged lid clinometer.

Fig. 2.1c The Chaix compass with the clinometer in the lid (compass courtesy of Topochaix Ltd., Paris).

side the structure to be measured, enabling the azimuth and plunge to be measured in one operation (see Section 2.3 for details). This allows rapid, accurate structural measurements to be taken, particularly of linear features.

2.2 Stereographic projections

The stereographic projection is a fundamental tool in structural geology and is used to represent 3D orientation data in a 2D graphical form. It is commonly used to solve problems involving the angular relationships of line and planes in

3D space. It cannot be used to solve problems involving the relative geographic positions of lines and planes.

It is beyond the scope of this Handbook to describe the construction and plotting of stereographic projections and the reader is referred to excellent texts on the plotting and manipulation of stereographic projections by Phillips (1971) or Ragan (1985). It is essential that before commencing any structural fieldwork the reader should be familiar with the plotting and simple manipulations of the stereographic projection.

2.2.1 Types of stereographic projection

Two types of stereographic projection may be used — the Wulff net or equal angle net (Fig. 2.2a) and the Schmidt or equal area net (Fig. 2.2b). The Wulff net is used to solve angular relationships, particularly where geometric constructions are made on the net, whereas the Schmidt net is used to solve angular relationships and to statistically evaluate orientation data using contoured stereographic projections. In this Handbook equal area lower hemisphere projections are used.

Where structural data are numerous it is appropriate to statistically evaluate them by contouring. In the field this can be easily done by using a counting net, the Kalsbeek net (Fig. 2.2c, p. 20) (see Ragan (1985) for details of counting and contouring techniques).

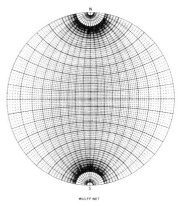

Fig. 2.2a The Wulff equal angle stereographic projection (reproduced with permission from Ragan, 1985).

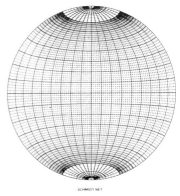

Fig. 2.2b The Schmidt equal area stereographic projection (reproduced with permission from Ragan, 1985).

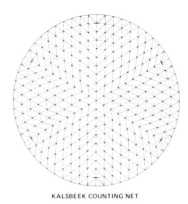

KALSBEEK COUNTING NET

Fig. 2.2c Kalsbeek counting net (reproduced with permission from Ragan, 1985). The equal area plot is placed on top of this net and the points falling into each hexagonal segment (1% of the total area of the net) are counted with the total posted at the centre point of each hexagonal segment. The totals at the centre points are then hand contoured in values of % per 1% area.

On stereographic projections:

1. planar structures plot as great circle lines but may also be commonly represented by poles (or normals) to the planes; these plot as points on the projection (Fig. 2.3).
2. linear structures plot as points.

2.3 How to measure structures

Most compasses can be adjusted to correct for the angular difference (declination) between magnetic north and geographic north. This correction should be made with reference to topographic maps of your area before you begin mapping, and should be recorded in your field notebook. The setting of your compass should also be checked period-

ically during the course of your mapping programme.

2.3.1 Conventions

Most geologists tend to record the attitude of planar structures as a strike and dip, e.g. **Strike 220°, Dip 45° SE.** Three measurements have to be made at each location: strike, dip and general dip direction, and ambiguous records are easily made when the directions of dip are often forgotten, and omitted. It is much safer and unambiguous to record *dip directions* for planar structures: e.g. **bedding 45°→130°,** i.e. a bedding dip of 45° in the direction 130° from north (Fig. 2.4), and for linear structures record the plunge, e.g. **plunge of minor fold axis 20°→120°,** i.e. a 20° plunge to the fold axis in the direction 120° from north (Fig. 2.4).

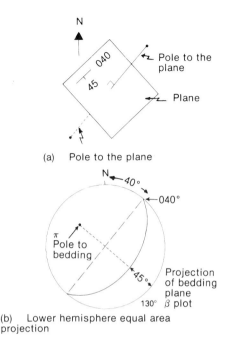

(a) Pole to the plane

(b) Lower hemisphere equal area projection

Fig. 2.3 (a) The pole to a plane. (b) stereographic projection of the plane and the lower hemisphere pole to the plane.

Structural data should be *consistently* recorded in the following formats: *Angles (measured from the horizontal) in 2 digits* e.g. 20°. *Azimuths (measured from north in a horizontal plane) in 3 digits e.g. 120°*, i.e. Fold plunge 20°→120° or dip direction of bedding 45°→130°. These sign conventions are unambiguous and definitive.

2.3.2 Methods for measuring—planar surfaces

Planar structures such as bedding, cleavage, schistosity, fold axial planes, fault planes, joints and veins are all measured in essentially the same way. The methods of measurement are illustrated using both conventional compasses such as the *Silva Ranger* 15 T and the *Freiberg* compass.

Bedding Plane

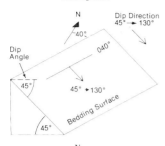

Fig. 2.4 (a) Bedding plane dipping 45° towards 130° (strike 040°, dip 45° SE) and showing the stereographic projection of the plane and the pole to the plane. (b) Minor fold axis plunging 20° towards 120° and showing the stereographic projection of the minor fold axis.

(a)
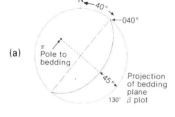

Plunging Minor Fold
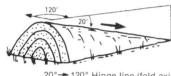

(b)
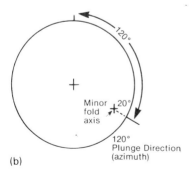

Method 1: Measurements with conventional compasses e.g. *Silva Ranger* 15 T. (Fig. 2.1a, 2.5a–d)

Method 1a: strike and dip method

1 Find the strike line (the horizontal line on the planar structure) by using the Silva compass as a clinometer and locating the direction of zero dip on the plane (Fig. 2.5a). Mark this line (the strike line) on the surface using a soft pencil (B or HB).

2 Measure the azimuth of this line (i.e. its direction from north) (Fig. 2.5b) — this is the strike of the plane. Record this angle, e.g. 220°.

3 Using your compass as a clinometer, place the edge at 90° to the strike line and measure the amount of maximum dip (Fig. 2.5c). Record this angle 45° and record the direction of dip SE.

The strike and dip of the plane is: **strike 220°, dip 45° SE.**

Note: if the surface is rough or uneven an average reading can be

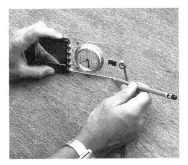

Fig. 2.5a Finding the strike line (zero dip line) on the bedding surface using the *Silva* 15TD-CL and marking the line with a soft pencil.

Fig. 2.5c Measuring the amount of dip (90° to the strike line) using the clinometer on the *Silva* 15TD-CL.

Fig. 2.5b Measuring the azimuth of the strike line.

Fig. 2.5d Use of the map board to carry out measurements on an irregular surface.

obtained by placing your map board on the surface and taking your measurements on this (Fig. 2.5d).

Method 1b: dip direction method

1   Using the compass as a clinometer find the direction of maximum dip on the plane (Fig.

2.6a). Mark this dip direction line on the surface and measure the angle of maximum dip. *Record this reading*—e.g. 45°.

2   Place your notebook or map-board along the dip direction line and holding it *vertical* measure the azimuth (direction) of the dip (Fig. 2.6b). *Record this reading*, e.g. 130°.

23

Fig. 2.6a Dip direction method using the Silva compass. Identify the position of maximum dip on the planar surface by using the clinometer on the compass. Mark this direction on the bedding surface and record the value of maximum dip.

Method 2: Measurements with the Freiberg compass

The Freiberg compass (Fig. 2.1b) allows rapid determination of the *dip directions* of a planar structure in one operation. Place the hinged lid of the compass against the surface to be measured and, keeping the base of the compass horizontal, (using the built-in spirit level) measure the azimuth of the dip direction e.g. 130° (Fig. 2.7). The dip can be read directly from the graduated hinge axis on the side of the compass e.g. 45°. *Record these readings.* The dip direction of the plane is **45°→130°**

Fig. 2.6b Determine the azimuth of the dip direction by placing your notebook along the line of maximum dip and keeping the notebook vertical, measure this direction.

The dip direction of the plane is **45°→130°**.

Note: always measure the direction or azimuth *looking down dip,* i.e. into the earth. Care must be exercised to measure the correct direction.

Fig. 2.7 Measure the amount and direction of dip of the bedding by placing the hinged lid of the Freiberg compass flat on the bedding plane.

Method 3: Measurements with conventional compasses or the Freiberg compass

Problems may arise when no suitable surface is available on which the compass can be placed, or the surface is too rough. In these situations the notebook or mapboard should be used to make a measurable plane parallel to the geological planar structure. This technique is illustrated in Fig. 2.8 where cleavage plane (Fig. 2.8a and b) or a fold axial plane (Fig. 2.8c) is measured by placing the map board parallel to it and then measuring the orientation of the mapboard using Methods 1a, 1b or 2 (the axial planes of folds are not commonly exposed as surfaces, so this method is often the only way to measure them).

Fig. 2.8b Measurement of the dip of a cleavage plane using a Silva compass and aligning the map board along the cleavage plane.

Fig. 2.8c Measurement of the orientation of the axial plane of a minor fold. Align the map board parallel to the fold axial plane and measure the dip direction using the Freiberg compass.

Fig. 2.8a Measurement of the strike of a cleavage plane by aligning the map board along the cleavage plane and measuring the azimuth (direction) of strike using the Silva compass.

Note: many students do not take enough readings in poorly-exposed areas because good planar surfaces are not exposed. Use of the map board to measure planar structures is essential in these situations.

Method 4: Sighting method

For moderate to steeply dipping planar structures it may be possible to measure the strike and dip by sighting along the compass (Barnes, 1981). This method is particularly useful when beds or outcrops do not form exposed planes convenient to measure. It is also useful for determining the average dip (or sheet dip) of a large outcrop or cliff exposure. In this method it is essential that you align your line of sight parallel to the stike of the beds or planar surface to be measured.

1 Align yourself so that your line of sight is parallel to the strike of the planar surface, (Fig. 2.9a).
2 Sight along your compass and measure the azimuth or bearing of the strike of the plane (Fig. 2.9a). *Record this azimuth* e.g. 150°.
3 Using the compass as a clinometer, align the edge of the compass with the dip of the planar surface (Fig. 2.9b) and measure the dip angle and the direction of dip, i.e. 60° SE. *Record these readings*.

The planar surface **strikes 150° and dips 60° SE**.

Fig. 2.9a Sighting method. Measure the strike of bedding by aligning the Silva compass parallel to the strike.

Fig. 2.9b Sighting method. Measurement of the angle of dip of the bedding using the clinometer of the Silva compass aligned parallel to the dip.

Remember that the techniques described above apply to the measurement of *any* planar surface, e.g. bedding, cleavage, schistosity, fold axial planes, joints, fault planes and veins.

2.3.3 Methods for measuring—linear structures

Linear structures (i.e. lines) include bedding/cleavage intersection lineations, mineral stretching lineations, minor fold axes or hinge lines, slickensides and crystal fibre structures. All linear structures are measured in the same way, either as a plunge (Fig. 2.10a) or as a pitch in a plane (Fig. 2.10b). See Fig. 2.10c for a stereographic projection of these measurements.

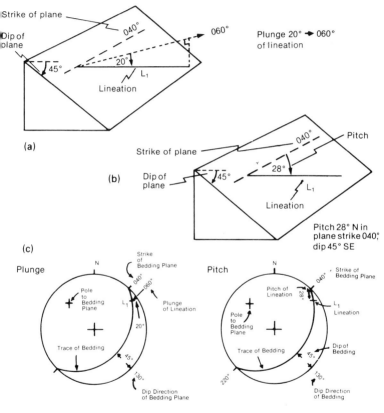

Fig. 2.10 (a) Measurement of the plunge of a lineation L_1 $20° \rightarrow 060°$ within the bedding plane. The angle $20°$ is measured in a vertical plane (azimuth $060°$) containing the lineation L_1.
(b) The pitch of a lineation L_1 is measured as $28°E$ from a line striking $40°$ east in the bedding plane.
(c) Lower hemisphere stereographic projections of plunge and pitch of the lineation L_1.

27

Method 1: Measurements with conventional compasses, e.g. Silva Ranger 15 T

Method 1a: Plunge of a lineation or fold axis

1   Place the edge of your mapboard or notebook along the linear structure to be measured (Fig. 2.11a and d). Keeping the mapboard or notebook vertical, measure the azimuth of the direction of the plunge (Fig. 2.11b and d). *Record this direction,* e.g. 060°.

2   Using the compass as a clinometer, measure the plunge of the linear structure by placing the edge of the compass along the structure (Fig. 2.11c and e). *Record this plunge,* e.g. 20°.

Plunge of the lineation 20°→060° (see Fig. 2.10a).

Method 1b: Pitch of a lineation

1   Find the strike of the plane as described in (Section 2.3.2) and Figs 2.5a and b. With your pencil inscribe the strike line on the planar surface so that it intersects the lineation L_1 (Fig. 2.12a). Measure the strike and dip of the plane (as described above in Section 2.3.2, Method 1a). *Record this data,* e.g. **strike 220° dip 45° SE.**

2   Lay your compass on the plane that you have measured, and use it as a protractor by aligning the edge of the compass along the strike line and rotate the calibrated outer ring until the reference arrow is parallel to the strike line. *Note this reading on the outer ring.* Now rotate this cali-

Fig. 2.11a   Measurement of a lineation L_1 — bedding/cleavage intersection in the bedding plane. Place the edge of your notebook along the lineation L_1.

Fig. 2.11b   Bring the notebook to the vertical and keeping it vertical, measure the azimuth of plunge of the lineation using the Silva compass.

Fig. 2.11c   Measure the amount of plunge of the lineation L_1 using the clinometer of the Silva compass.

Fig. 2.11d Measurement of the plunge of a minor fold axis. Place the map board along the fold axis and keeping it vertical, measure the azimuth (direction) of the plunge.

Fig. 2.11e Measure the amount of plunge of the fold axis by placing the edge of the Silva compass on the fold axis and using the clinometer.

Fig. 2.12a Measurement of the pitch of the bedding/cleavage intersection lineation L_1 on the bedding plane using the Silva compass. Mark the strike line on the bedding plane and determine the strike and dip of the bedding as in Fig. 2.5.

Fig. 2.12b Using the Silva compass as a protractor, place it on the bedding plane and measure the angle between the strike of the bedding and the lineation L_1 on the bedding plane (see text for details).

brated outer ring until the reference arrow is parallel to the lineation (Fig. 2.12b) and *note this new reading and the direction of the pitch of the line*. The difference between the two readings is the pitch of the L_1 in the plane. *Record this data,* e.g., the pitch of the line is **28° North** in the plane which has a strike of **220°** and dip of **45° SE** (see Fig. 2.10b).

Note: The method of measuring pitches is most useful on steeply dipping planes where accurate determinations of plunge can be difficult, and when the lineation itself is steeply plunging, making accurate plunge determinations difficult and resulting in large errors in azimuth. Reference to Fig. 2.10c will remind you that using a stereonet plunge can be obtained from a measure of pitch, and vice-versa.

Method 2: Measurements with the Freiberg compass

Place the side-edge of the hinged compass lid along the linear structure to be measured and then, keeping the base of the compass horizontal, read off the azimuth of the plunge direction. Then measure the amount of plunge on the calibrated hinge pin. *Record this data*, e.g. **20°→060°**. This method is illustrated in Fig. 2.13 for the measurement of a minor fold axis.

Note: In some cases it may not be possible to align the compass directly on the lineation. In these cases the linear structure may be extended by placing a pencil parallel to the lineation (i.e. in a weathered out fold hinge) and then measuring the plunge of the pencil (Fig. 2.14).

2.4 The field map and aerial photographs

The field map, aerial photographs and field notebook are the most important records of your field observations. Take great care of them. They must be:

1 *labelled with your name and address,*
2 *neatly and carefully drafted and legible,*
3 *fully annotated with legends, symbols and scales, and contain all the necessary location information,*
4 *completed whilst you are in the field.*

The importance of completing your map whilst in the field cannot be overemphasised, as it is only then that you can carry out an ongoing

Fig. 2.13 Measurement of the plunge of a minor fold axis by placing the edge of the Freiberg compass on the hinge line.

Fig. 2.14 Measurement of the plunge of a minor fold axis by using a pencil to extend the axis.

interpretation, construct cross-sections and identify key and problem areas that warrant further work.

2.4.1 Styles of mapping

The style of mapping used is largely controlled by the map scale, the degree of structural complexity and the degree of exposure. If good detailed topographic maps are available then field data may be plotted directly onto these. Where insufficient detail is shown on the topographic map, aerial photographs must be used in the field for accurate location of outcrops, and to map in lithological boundaries and structural trends. The data is subsequently transferred onto the base map (Barnes, 1981). Barnes (1981) reviews various mapping styles, and these are summarised below.

1 *Traversing* is used mainly for regional mapping at scales of 1:250,000 to 1:50,000.

2 *Contact mapping* is used mainly for more detailed mapping at scales of 1:50,000 to 1:15,000.

3 *Exposure mapping* is detailed mapping during which the location and size of each outcrop are recorded, usually at scales of 1:15,000 to 1:1000.

4 *Baseline mapping* involves detailed mapping using a measured baseline (or compass and pacing) at scales of 1:10,000 to 1:500.

5 *Grid mapping or plane table mapping* are techniques used for detailed exposure mapping at scales of 1:1000 to 1:1.

Traversing: In structurally complex areas this is the best method of quickly establishing basic stratigraphic and structural relationships. It can be achieved by traversing perpendicular to the strike of the dominant structural trend and by constructing sketch cross-sections in the field. Fig. 2.15 shows a sketch section through a simple antiformal structure (note the use of bedding/cleavage relationships to locate the hinge region of the antiform).

Contact Mapping: This technique involves following lithological or structural contacts in order to establish 3D structural relationships. For example, it may be necessary to determine if a fault cuts up or down stratigraphy or to establish the outcrop pattern in a polyphase deformation terrane.

Exposure Mapping: This method is essential for detailed structural studies in complexly deformed areas. In particular it is used to establish structurally homogeneous sub-areas and to establish interference relationships in areas of complex folding. An example of exposure mapping is shown in Fig. 2.16.

Baseline, grid and plane table mapping: Detailed mapping using these techniques is essential in establishing detailed relationships in one outcrop or in a group of closely-spaced outcrops. Key structural relationships are illustrated by these methods. An example of baseline mapping is shown in Fig. 2.17.

2.4.2 Map scales

A *detailed structural map* may be produced at any scale from 1:250,000 to 1:1. The same types of structural data should be collected at every locality, *irrespective* of the scale at which you are mapping. Failure to measure all available structural elements may seriously hinder your future interpretations. Avoid using base maps which have been excessively enlarged from a large scale topographic map: they are no more accurate than the maps from which they come.

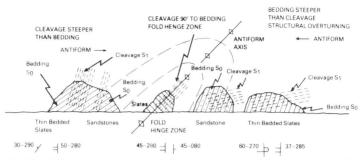

Fig 2.15 Field sketch through an antiform, illustrating the use of bedding/cleavage relationships to determine the position of the hinge of the fold. Structural data are plotted at the base of the section so that the section can be related to the map more readily. Symbols as in Table 2.1 (p. 42).

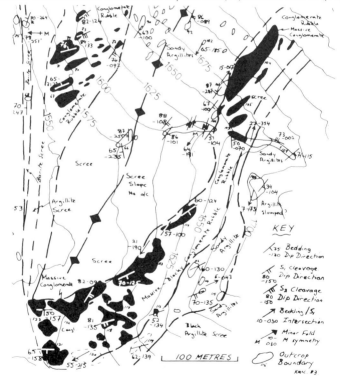

Fig. 2.16 An example of exposure mapping in an area with two cleavages developed.

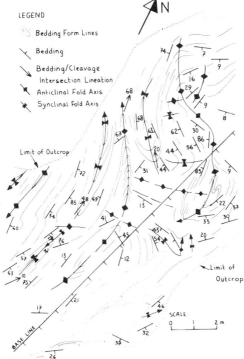

Fig. 2.17 An example of detailed baseline mapping with the locations measured off from the baseline (e.g. Barnes, 1981).

2.4.3 Aerial photographs

In many situations, you will map directly onto overlays on aerial photographs (Fig. 2.18). Provided the central region of the aerial photograph is used, then problems due to excessive distortion can be minimised. Structural data, outcrop boundaries, formation boundaries, major fold axes and fault traces, and locality numbers are plotted directly onto the photograph overlay (Fig. 2.18). These are subsequently transferred onto your base maps using the techniques outlined in Barnes (1981). Aerial photographs are particularly useful for accurate outcrop location, mapping lithological boundaries, and identifying and mapping structural features. In many areas the structural grain can be readily seen on the aerial photograph but difficult to pick out on the ground. Use of aerial photographs is a skill that is acquired through practice and patience. Great care needs to be taken in accurately locating yourself and becoming accustomed to the scale of the photograph.

Fig. 2.18 Overlay of an aerial photograph, showing the exposures mapped, and traces of lithological boundaries and structures plotted on the overlay.

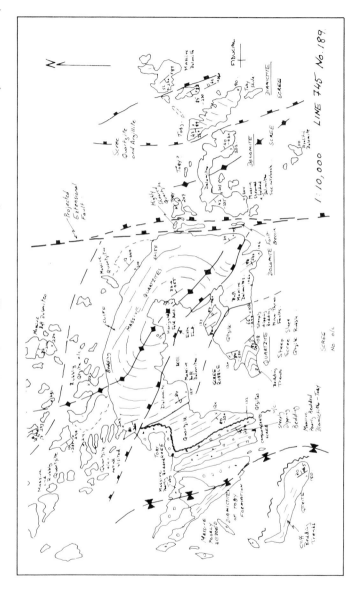

2.5 Field notebook

As with your field map, your field notebook is an important record of your field observations. It must be neat, legible, clearly written and well illustrated. Take care of it!

The field notebook is a record of activity and must contain the appropriate location and reference data, so that together with your field map it can be *interpreted by someone else*. Do not adopt the habit of treating it as your own personal record: when you are employed your notebook will belong to your employer. Do not use your own shorthand and hieroglyphics. You should record as much detail as possible and fully illustrate your notebook with sketches (3D if possible), interpretative sections and maps. Nothing is more frustrating upon returning to the laboratory than finding that your field notes are incomplete.

The key to producing good field notes is keen, careful observation and systematic recording. The following recording procedure should be adopted *at every locality:*

1 Date, time and location of your observations. Use grid references and aerial photograph reference number where appropriate.
2 Resume of your mapping method — e.g. traverse up Steamboat Creek from road bridge at Km 14.
3 Outcrop locality number — to be also marked on your field map. Include a brief summary of the outcrop characteristics — size and general nature of the exposure.
4 Record the lithological characteristics (for details see Barnes, 1981, Tucker, 1982, Fry, 1984, and Thorpe and Brown, 1985).
5 Record the structural characteristics — descriptions and measurements (for details see following chapters).
6 Sketch the outcrop and structural relationships.
7 Record the collection of samples and the photographs taken.
8 Interpret the outcrop in terms of the regional setting and draw sketches of the structural relationships.

2.5.1 Examples of field notebooks

Examples of field notebooks are shown in Figs. 2.19 and 2.20. Fig. 2.19 shows the information collected at one small outcrop in the course of a regional mapping programme, whereas Fig. 2.20 shows the information gathered during a detailed analysis of an outcrop that exhibits polyphase deformation. In the latter case, a large amount of descriptive and orientation data has been collected. In both examples, more than one reading for each structural element has been taken. An average or representative reading for each structural element is then plotted on the field map. Note that your field notebook should always contain much more structural data than can be plotted on the field map.

Fig. 2.19 Example of a field notebook showing data collected at a small outcrop during the course of a regional mapping programme.

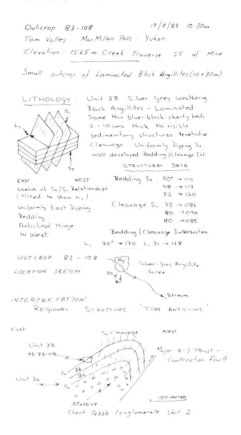

Outcrop 83-108   17/8/83 10 30am
Tom Valley . MacMillan Pass : Yukon .
Elevation : 1565m. Creek Traverse   SE of Mine

Small outcrop of Laminated Black Argillites (10 x 30m)

LITHOLOGY : Unit 3B . Silver Grey Weathering
             Black Argillites – Laminated
             Some thin blue-black cherty beds
             2 - 10 cms thick . No visible
             sedimentary structures Penetrative
             Cleavage . Uniformly Dipping So .
             Well developed Bedding/Cleavage Int.

                    STRUCTURAL DATA

EAST   WEST   Bedding So . 50° → 110
Sketch of So/S1 Relationships   48 → 117
(Tilted to show L1)   52 → 120
Uniformly East Dipping   Cleavage S1 75 → 086
Bedding .   80 → 090
Anticlinal Hinge   80 → 085
to West   Bedding/Cleavage Intersection
                 L1 30° → 170 L1 31 → 168

OUTCROP   83 - 108
LOCATION SKETCH.   Silver-Grey Argillite
                                      Scree
                                      Stream.

INTERPRETATION.
        REGIONAL STRUCTURE 'TOM ANTICLINE'

East   S1 Cleavage   West
   Unit 3B   Major N-S Thrust –
   O/c 83-108   Contraction Fault
   So
Unit 3a
   So   150 METRES
              Massive
              Chert Pebble Conglomerate Unit 2

Fig 2.20 (pp. 37–40) Pages from a field notebook showing data collected during detailed structural observations in a complex polyphase terrane. Key outcrops must be examined in detail in order to determine the structural history. In this case an hour was spent examining the large outcrop, sketching the structural relationships and taking several measurements of each structural feature. The following pages outline the methodology used: **1** Identify and describe the lithologies. **2** Identify and describe the structures. **3** Measurements of planar structures S_0, S_1, S_2, S_3. **4** Measurements of the lineations L_1, L_2, L_3, etc. **5** Measurement of minor fold axes and axial surfaces. **6** Detailed sketches of structural relationships.

36

1.

O/C No 17   RHOSCOLN   ANGLESEY   9-11-84

Location: Coastal outcrop in 30m high rocky cliffs.

Lithology: Pelite Unit with thin (<1cm - 3cm) psammite
interbeds, Green; Buff Weathering Strongly
folded into kink like folds. Fine-grained Pelites
Chloritic. Contains abundant folded quartz
veins

Thickness: 15 m thick sandwiched between two massive
quartzite units. _Sheet Dip_ towards 120° ~25° ✳

Structure: S₀ Bedding — Footwall and Hangingwall Massive
quartzite units thickly bedded. No observed sedimentary
structures.

Pelite unit contains thin disrupted psammite
beds - lenticular which are transposed into S₁

Foliations: Cleavage (S₁?) This foliation strongly folded into
kink like folds with pressure solution axial planar
fabric (S₂?). Lenticular quartz veins parallel
to S₁ and folded by kink-like folds

Structural Data: Bedding S₀. Footwall Quartzites.

S₀ 42° →116°

S₀ 44° →118°

S₀ 44° →124°

Note: This is a sheet dip (enveloping surface) in this unit.

1 Outcrop location and description and identification of the lithologies. In this outcrop two
contrasting lithologies occur — massive quartzites and highly contorted pelitic units. Record the
observations of structural features and measurements of the *sheet dip* of the bedding (Fig. 3.1).

O/c No 17 continued   STRUCTURAL DATA CONT'D

Foliations :   Footwall Quartzites : Well developed penetrative

             S_2 foliation — slightly anastomosing on bedding surface

S_2 $61° \rightarrow 337°$ Note : anastomosing character may be

S_2 $67° \rightarrow 343°$ due to weak development of S_1

S_2 $68° \rightarrow 342°$ Foliation

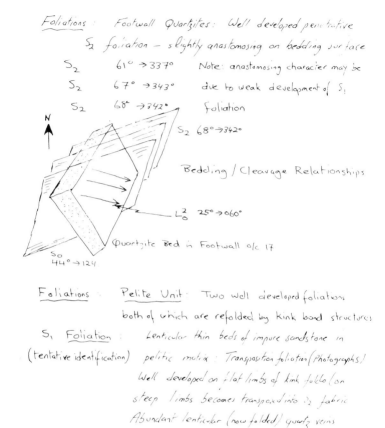

S_2 $68° \rightarrow 342°$

Bedding / Cleavage Relationships

L_0^2 $25° \rightarrow 060°$

Quartzite Bed in Footwall o/c 17

S_0
$44° \rightarrow 124$

Foliations :   Pelite Unit : Two well developed foliations

             both of which are refolded by kink band structures

S_1 Foliation :   Lenticular thin beds of impure sandstone in

(tentative identification)   pelitic matrix : Transposition foliation (Photographs)

             Well developed on flat limbs of kink folds (on

             steep limbs becomes transposed into S_2 fabric

             Abundant lenticular (now folded) quartz veins

             parallel to S_1 fabric

Flat ⎫ S_1 $31° \rightarrow 024°$ S_1 $30° \rightarrow 044$
Limb ⎭ S_1 $33° \rightarrow 054°$ (Note : Variable S_1 on F_3 fold limb)

Fig. 2.20a (continued) **2** Description and record of the orientation data for the S_2 foliation and the L_0^2 lineation in the quartzite unit. Description and record of the orientations of the S_1 foliation in the pelitic horizons.

3.

o/c 17 Continued. RHOSCOLYN 9/11/84

Foliations: S_1 in Pelitic Unit

Steep Limb	S_1 77° → 325°	S_1 68 → 338°	Note: here S_1 sub-‖ to S_2.
	S_1 70° → 354°	S_1 72 → 343°	

S_2 Foliation: Pressure solution style cleavage on limbs of F_2 crenulation and kink folds.

S_2: 60° → 358°	S_2: 50° → 004°	S_2: 50° → 358°
S_2: 60° → 350°	S_2: 68° → 348°	S_2: 55° → 357°

S_3 IN KINK BAND

F3 Reverse Kink 76° → 153° F_3 AP

F3 Reverse Kink 59° → 142° F_3 AP.

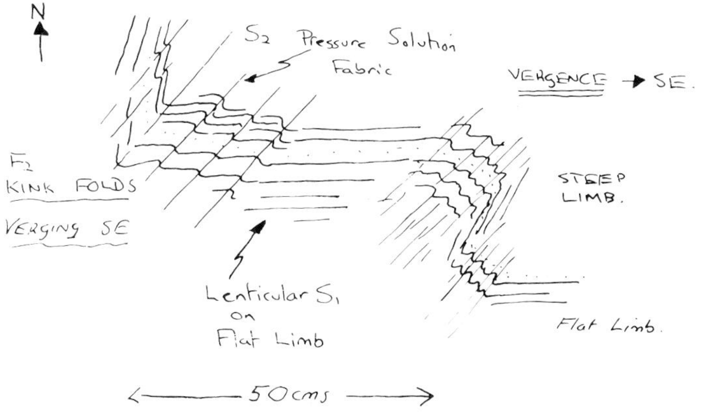

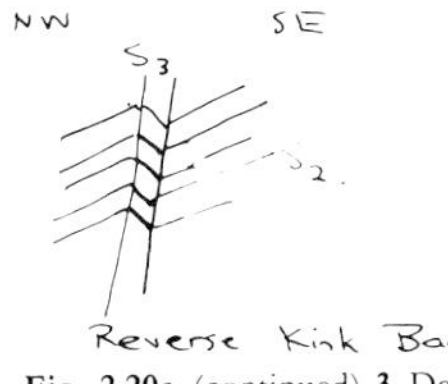

Fig. 2.20a (continued) **3** Description and record of orientation data for the S_2 foliation in the pelitic units. Sketches and descriptions of the F_3 kink bands and the S_3 foliation surface. The sketches of outcrops show both their geographic orientation and vergence. (Note that space does not permit examples of lineation measurements to be included in this figure.)

Fig. 2.20a (continued) **4** Detailed field sketch showing the structural relationships in the outcrop described in the previous pages.

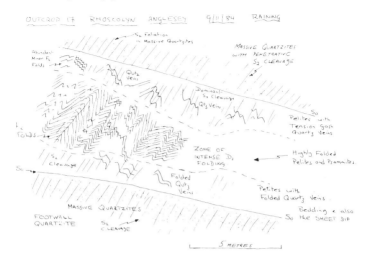

Fig. 2.20b Field map entry for the outcrop described in (a) above. Note that only representative structural data and the overall sheet dip have been plotted.

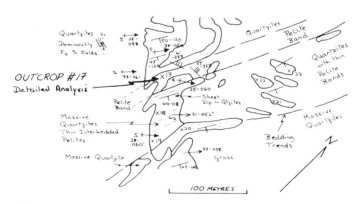

2.6 Map symbols

Map symbols must be clear and unambiguous. For most map areas, representative data of all structural elements should be plotted at each locality on the field map, with the possible exception of joints or veins. Unless required, the plotting of joints may serve no useful purpose and will only clutter the map. Various coloured inks may be used to indicate different geological features — e.g. bedding or lithological features — black; structural features — red; outcrop boundaries — blue or green; geomorphological features (including alluvium and drift) — brown. Unfortunately, coloured inks do not reproduce well and black ink may be preferred for all symbols.

Remember, geological data from outcrops should always be plotted in preference to geomorphological features.

Plotting symbols

The following are emphasised:
1 On the map symbols should be plotted directly at the map location for the outcrop from which the measurements were taken, and *not* left 'floating' in areas of the map where no exposure is recorded.
2 The azimuths of all orientation data should also be plotted on the field map, together with dip and plunge data (Fig. 2.16). This enables the accuracy of the data to be checked and allows data to be extracted from the map for further analysis.

3 Data should be plotted on the map either using a protractor or using the Silva compass directly (Barnes, 1981, p. 60).

4 *Plotting should always be done in the field.*

Table 2.1 gives a list of suggested map symbols that can be used in the field.

2.7 Oriented samples

It may be necessary to collect oriented samples for (*a*) the analysis of preferred orientations, (*b*) the investigation of the superposition of foliations or (*c*) the investigation of relationships between metamorphic mineral growth and tectonic fabrics, or (*d*) strain determinations (Appendix III).

2.7.1 *Sample collection*

1 Select the sample to be collected (pieces bounded by joint surfaces are the easiest to collect and are less likely to break upon extraction).
2 Measure and record the structural elements associated with the sample and the outcrop.
3 Select an appropriate reference plane on the surface of the sample — this will generally be either bedding (S_0), a foliation surface (S_1), as shown in Fig. 2.21, or a joint surface.

Table 2.1

MAP SYMBOLS

GENERAL SYMBOLS

STRUCTURE	Observed Lithological Boundary	Boundary Position Uncertain	Boundary Position Infered	Outline of Outcrop with Locality Number	Younging Direction	Morain Ridge	Break of Slope	Alluvium Boundary
MAP SYMBOL				126	or			

PLANAR STRUCTURES

STRUCTURE	Bedding	Overturned Bedding	First Cleavage	Second Cleavage	Minor Fold Axial Plane (phase 1)	Minor Fold Axial Plane (phase 2)	Joints (set 1)	Joints (set 2)
TECTONIC SYMBOL	S_0	S_0	S_1	S_2	F_1 AP	F_2 AP	J_1	J_2
MAP SYMBOL	30—110	30—110	60—110	45—115	60—110	45—115	55—115	30—115

FOLD STRUCTURES

STRUCTURE	Syncline with Dip Direction of Axial Plane	Anticline with Amount and Direction of Plunge	Synform	Antiform	Overturned Syncline	Overturned Anticline
MAP SYMBOL	45—150	15—041				

FAULT STRUCTURES

STRUCTURE	Extensional Fault High Angle	Extensional Fault Low Angle	Contractional Fault High Angle	Contractional Fault Low Angle	Wrench Fault	Shear Zone
TECTONIC SYMBOL	E Fault / N Fault	E Fault	C Fault / R Fault	C Fault / T Fault	W Fault	S Z
MAP SYMBOL	70—150	20—145	65—165	15—160	90—160	90—155

LINEAR STRUCTURES

STRUCTURE	Bedding/ Cleavage(S1) Intersection	Cleavage(S1) Cleavage(S2) Intersection	Mineral Stretching Lineation on S1	Minor Fold Axis (phase 1)	Minor Fold Axis (phase 2)	M Fold Axis	Z Fold Axis	S Fold Axis
TECTONIC SYMBOL	L_1	L_2	ML	MF_1 A	MF_2 A	M-MF_1 A	Z-MF_1 A	S-MF_1 A
MAP SYMBOL	10—050	15—055	15—060 ML	45—050	35—051	50—050 M	40—045 Z	56—045 S

42

4 Measure the orientation of this surface, and with a *waterproof* marking pen inscribe the strike and dip on the reference surface, (Fig. 2.21). Mark on the *top* of the specimen and the specimen number. Record the data in your notebook and draw a *sketch* of the specimen and its structural relationships.

5 Collect the specimen and bag it, fully labelling the sample bag.

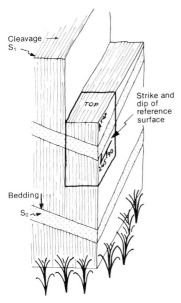

Fig. 2.21 Collection of an oriented specimen. Reference marks on the specimen indicate the way-up and the strike and dip of the reference surface.

2.8 Photography

Photography is an important method of recording geological information. However, it is no substitute for good, detailed field sketches. It is often very difficult to distinguish or interpret structures from a photograph without a good field sketch. The following points are therefore emphasised.

1 A good 35 mm camera (preferably a single lens reflex) is essential. Colour slide film (easily projected to a large size) or black and white film (easily enlarged) are preferred. Close-up lenses or attachments may be useful.

2 *Always draw a sketch* of the area or structure being photographed (e.g. Fig. 2.22).

3 Record the information about the photograph in your field notebook, in particular noting the direction of view.

4 Where possible always include an easily recognised scale in the photograph.

5 Fill the frame with the structure being photographed. Many students do not get close enough to the subject and all detail is lost in the photograph.

6 In some instances stereoscopic photographs may be taken to aid later interpretation. In this case two photographs are taken approximately 1.5 m apart in a line parallel to the exposure being photographed. Sixty per cent

43

overlap is required and the resultant pair of photographs can be viewed stereoscopically for analysis. This technique is particularly useful for fracture studies.

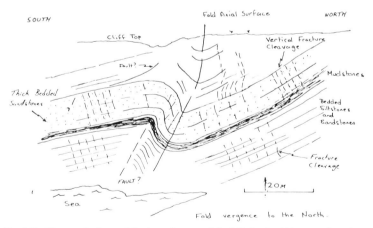

Fig. 2.22 Photograph of an anticline in sandstones and the field sketch which was made at the same time as the photograph was taken.

3
Fold structures

Here, attention is focused upon fold geometries, nomenclature, description and measurement. Correct description and measurements of folded surfaces gives valuable information on the geometry and mechanics of folding.

3.1 Basic fold nomenclature

Basic fold nomenclature is outlined in Fig. 3.1. Care must be taken to distinguish terms which describe the orientation of the folded surface (e.g. hinge line) from those which describe aspects of its spatial orientation (e.g. fold crest line). The following are emphasised:

1 The fold *hinge line or fold axis* is the line of maximum curvature on the folded surface (Fig. 3.1a).
2 The fold *axial plane* is the plane containing the hinge lines within one particular fold (Fig. 3.1a) (Note: many fold axial planes are *curved* (e.g. Fig. 2.22) — not *planar*, and the term *axial surface* is preferable.)
3 A fold is *symmetric* if the limbs either side of the axial plane are of equal length, and the fold is *asymmetric* if they are not. (Fig. 3.1b).
4 The fold *wavelength* is the distance

between adjacent hinge lines or inflexion points (Fig. 3.1b).
5 A fold is *cylindrical* if it has the same shape in the profile plane at all points along the fold axis (Fig. 3.1b). A *non-cylindrical fold* has a varying profile shape along the fold axis (Section 3.2.2 and Fig. 3.7).
6 A *fold train* is a series of folds within a particular unit or series of units (Fig. 3.1b).
7 The concept of *enveloping surface* for a series of folds (such as Fig. 3.1b) is extremely important. The enveloping surface is drawn tangential to the fold hinges (or through the inflexion points) (Fig. 3.1b). This is particularly important when mapping areas with abundant small amplitude, short wavelength folds which obscure the overall *sheet dip* (i.e. enveloping surface) of a unit.

3.2 Fold types

Most texts on structural geology describe fold types in detail (e.g. Ramsay, 1967; Hobbs *et al.,* 1976). Here we are concerned with the description of folds as seen in the profile section (i.e. the section perpendicular to the fold axis (Fig. 3.2).

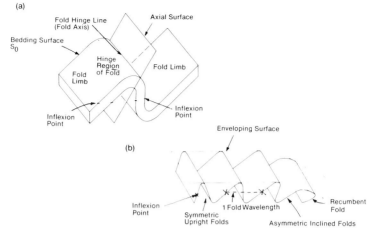

Fig. 3.1 Basic fold nomenclature and fold architecture. (*a*) A single inclined fold pair showing basic nomenclature. (*b*) A fold train consisting of upright folds that pass into inclined and then recumbent folds. Note the concept of the enveloping surface, which is either tangential to the fold hinge lines, or passes through the inflexion points (points of maximum slope) on the folded layers.

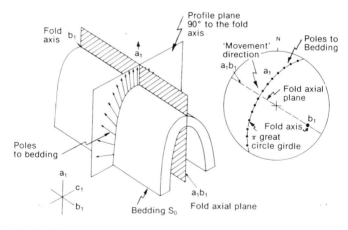

Regional Tectonic Axes

Fig. 3.2 Fold architecture showing the profile plane; fold axial plane and relationship of both to the regional tectonic axes: 'a_1' indicates the direction of tectonic transport, 'b_1' is parallel to the fold axis, 'c_1' is normal to a_1 and b_1; and the stereographic projection of data from the fold. Note the use of poles to bedding to determine the fold axis b_1.

46

Common types of fold (*as described in the profile plane*) are:

1 *Parallel* folds (Fig. 3.3a) Orthogonal thickness (i.e. thickness perpendicular to the folded surface) is constant.

Fig. 3.3a Parallel folds. Folded quartzo-feldspathic layer in amphibolites. Field of view *ca.* 2·5 m.

2 *Similar* folds (Fig. 3.3b). Thickness parallel to the axial plane is constant.

Fig. 3.3b Similar folds (thickness parallel to the axial plane is constant) in deformed psammitic schists. Note that these folds are also harmonic in that there is continuity along the axial planes.

3 *Harmonic* folds (Fig. 3.3b). Axial planes are continuous across a number of layers.

4 *Disharmonic* folds (Fig. 3.3c). Axial planes are not continuous from one layer to the next.

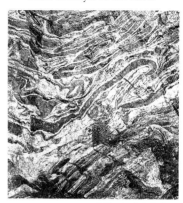

Fig. 3.3c Disharmonic folds of sphalerite layers in galena. There is no continuity along the axial planes between layers. Field of view *ca.* 1 m.

5 *Intrafolial* folds (Fig. 3.3d). Folds contained within the layering or foliation.

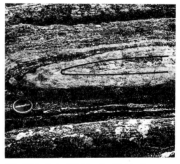

Fig. 3.3d Intrafolial isoclinal fold of a quartzo-feldspathic layer in gneisses. The fold is contained within the gneissic foliation.

47

6   *Ptygmatic* or elastica folds (Fig. 3.3e). Tight folds in which the fold limbs fold back on themselves so that the angle between the fold limbs at the hinge of the fold has a negative value.

7   *Chevron* folds (Fig. 3.3f). Angular folds with planar limbs and sharp hinges.

8   *Isoclinal* folds (Fig. 3.3d). Folds in which the limbs are strictly *parallel*.

9   *Polyclinal* folds (Fig. 3.3g). Folds with more than one axial plane, e.g. box folds or conjugate kink bands.

10 *Kink bands* (Fig. 3.3h). Sharp angular folds bounded by planar surfaces.

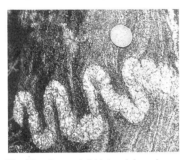

Fig. 3.3e Ptygmatic folds in a deformed pegmatitic vein in gneisses.

Fig. 3.3g Polyclinal fold in deformed mylonites and showing several axial planes.

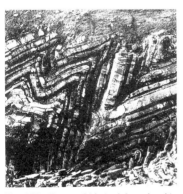

Fig. 3.3f Chevron folds in deformed turbidites. Note the sharp hinges and planar limbs of the folds. Field of view *ca.* 15 m.

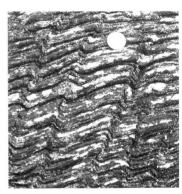

Fig. 3.3h Asymmetric reverse kink bands in a deformed schist.

3.3 Analysis and classification of folds

Folds are classified by:
1. The attitude of their hinge lines.
2. The attitude of their axial surfaces.
3. Their interlimb angles.
4. Their symmetry—i.e. length ratios of fold limbs.
5. Shapes of the folded layers.
6. Their degree of cylindricity.

3.3.1 Two dimensional classification methods

Tightness of folds: The *interlimb angle* measured between *inflexion points* (Fig. 3.1) can be used to measure the tightness of folds (Table 3.1).

Orientation of Folds. The orientation of a fold is completely defined by the direction of closure and the attitudes of the hinge line (i.e. fold axis) and hinge surface (i.e. axial plane). This classification nomenclature is outlined in Table 3.2.

Table 3.1 Terms used to describe the tightness of folds

	Interlimb angles	Fold tightness
	180°–120°	Gentle
	120°–70°	Open
	70°–30°	Close
	30°–0°	Tight
	0°	Isoclinal
	less than 0°	Elasticas
	-ve angle	or Ptygmatic

Table 3.2 Terms describing the attitude of folds

Dip of the fold axial surface or plunge of the fold axis	Dip of hinge surface (i.e. attitude of axial plane)	Plunge of hinge line (i.e. attitude of fold axis)
0°	Recumbent fold	Horizontal fold
1°–10°	Recumbent fold	Sub-horizontal fold
10°–30°	Gently inclined fold	Gently plunging fold
30°–60°	Moderately inclined fold	Moderately plunging fold
60°–80°	Steeply inclined fold	Steeply plunging fold
80°–89°	Upright fold	Sub-vertical fold
90°	Upright fold	Vertical fold

Note: Two attitudes are used to describe the orientation of a fold, (a) the *plunge* of the *hinge-line* or *fold axis* and (b) the *dip* of the *axial plane*. Both are required to correctly describe the fold attitude.

Shape of Folds: Folds may be classified according to the shape either of individual surfaces or of folded layers.

(a) *Dip isogons* The shape of the folded layers is the feature most usually recorded and is quantified by the use of dip isogons. First find a profile plane through a fold, then draw lines which join points of equal dip through the stack of folded layers. The reference datum is the tangent which passes through the hinge point of the folded surface. Three basic classes of dip isogon patterns are found (Fig. 3.4):

Class 1 folds: Folds with convergent dip isogons.

Class 2 folds: Folds with parallel dip isogons: similar folds.

Class 3 folds: Folds with divergent dip isogons.

Note: Convergence and divergence are measured going from the outer arc to the inner arc of the fold. Dip isogons may be used to construct cross-sections and are particularly useful in metamorphic terranes where bedding thicknesses change around a fold.

(b) *Fourier or harmonic analysis* The description of fold shapes may be

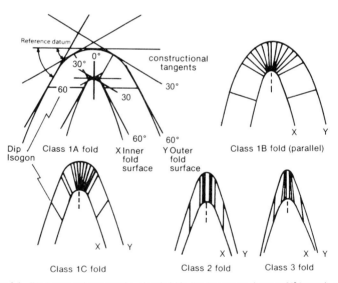

Fig. 3.4 Dip isogon technique used to classify folds. Dip isogons are drawn at 30° intervals and join points of equal dip on the upper and lower fold surfaces. Tangents used in the construction are shown on the first fold.

further quantified by considering the harmonic waveforms, which when combined, form the fold shape. This technique was devised by Huddleston (1973) and a simplified version may be applied by visual inspection using the chart of 30 idealised fold shapes shown in Fig. 3.5. Six basic fold shapes are recognised: A — *Box folds,* B — *Curved double hinged folds,* C — *Semi-ellipses,* D — *Parabolas,* E — *Semi-chevron* and F — *Chevron folds.* This chart (Fig. 3.5) is also graduated in amplitudes from 1 to 5 and hence can be used in the field to classify folded surfaces.

3.3.2 Three-dimensional classification methods

A different classification to that described for 2D shapes is required when folds vary in character along their length, i.e. when their shape as seen in the profile plane changes along the hinge of the fold. Williams and Chapman (1979) have constructed a simple classification diagram for folds, based upon the measurement of interlimb angle α, the angle made by the curving hinge line β, and the angle made by the

FOLD SHAPE

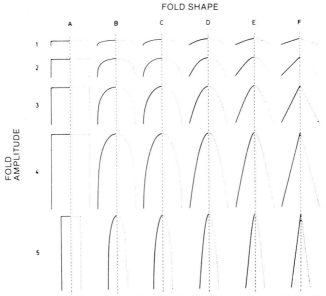

Fig. 3.5 Visual harmonic analysis. Field chart for Fourier shape classification of folds. The fold amplitude is measured on a scale 1–5; the fold shape is measured in columns A to F. A = *box folds,* B = *intermediate* between *box folds* and *semi-ellipses,* C = *semi-ellipses,* D = *parabolas,* E = *intermediate* between *sine waves* and *chevron bolds,* F = *chevron folds.* (reproduced with permission from Huddleston, 1973).

curving fold axial surface γ (Fig 3.6). The triangular PQR diagram has planes, cylindrical isoclines and isoclinal domes at its corners (Fig. 3.7). This classification may be used in the field.

3.4 Symmetries of parasitic minor folds

Fold symmetry is characterised by considering the relative lengths and attitudes of the *long — short — long*

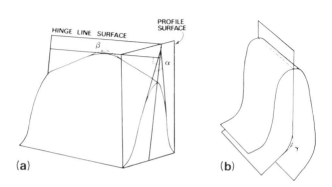

Fig. 3.6 Geometric description of non-cylindrical folds. (*a*) α = the interlimb angle, β = hinge angle, (*b*) γ = hinge line surface angle. (reproduced with permission from Williams and Chapman, 1979).

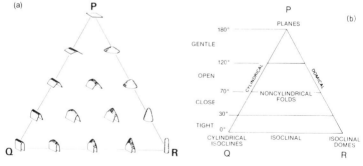

Fig. 3.7 Non-cylindrical fold shapes plotted on a PQR diagram. (*a*) Diagrammatic representation of fold shapes. (*b*) Descriptive terminology (reproduced with permission from Williams and Chapman, 1979).

limbs of a minor fold (Fig. 3.8a). Z, S, and M symmetry minor folds can be identified (Fig. 3.8a). M folds have fold limbs of equal length, and hence have no asymmetry.

Fold symmetry is always determined by looking down *the plunge of the minor fold.* (Note that a minor fold with Z asymmetry looking down plunge would have an S sense of asymmetry when viewed up plunge: so, always look down plunge.)

In a large fold structure the sense of asymmetry of minor parasitic folds will vary systematically around the structure and as such can be used to determine the position of an outcrop within the larger fold structure (Fig. 3.8b). S and Z folds are found on the limbs of major folds, whereas M folds indicate the hinge region of a large fold structure (Fig. 3.8b). Systematic determination and recording of minor fold asymmetries is a powerful tool for identifying major fold structures and should always be carried out during the course of your mapping programme. Minor fold symmetries must always be plotted on your map (Fig. 3.8c).

3.5 Vergence

Vergence is a term used to indicate the direction of movement and rotation that occurred during deformation. The concept of vergence may be applied to asymmetric folds and to the relationship of one cleavage to another, although this is a complex procedure and is not described in this Handbook (for details see Bell, 1981). Vergence is extremely useful in complex deformed terranes, and you should always attempt to evaluate vergence relationships when mapping.

3.5.1 Fold vergence

Vergence of *asymmetric* folds is defined as the horizontal direction of movement of the *upper* component of a fold (measured in the profile plane, e.g. Fig. 3.9). Minor folds of S and Z asymmetry may have the same vergence, M folds in the hinge region of a major fold have neutral

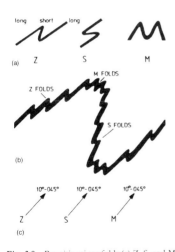

Fig. 3.8 Parasitic minor folds (*a*) Z, S, and M symmetry minor folds as viewed looking down the fold plunge; (*b*) Symmetry of parasitic minor folds around major fold structures; (*c*) Map symbols for minor folds and their symmetry.

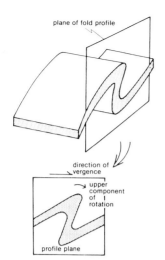

Fig. 3.9 Definition of fold vergence (reproduced with permission from Bell, 1981). For examples, see Figs. 2.20 and 2.22.

vergence, and vertically plunging folds have either sinistral or dextral vergence (Fig. 3.10). The principal use of minor fold vergence is to locate major fold axial surfaces. In simple geometries, minor folds change vergence across major fold surfaces (Fig. 3.11).

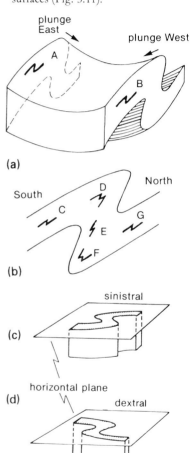

Fig. 3.10 Minor fold vergence relationships around a major fold structure. (a) plunge reversals lead to reversals of sense of asymmetry, but not of vergence. Minor fold A has an S sense of asymmetry (looking down plunge to the east) whereas minor fold B has a Z sense of asymmetry (looking down plunge to the west). Both minor folds have vergence to the north towards the hinge of the anticlinal structure; (b) minor fold vergence used to locate the major fold axes. In cross-section the minor folds C and E verge towards the hinge of the anticline, whereas folds D and F have neutral vergence at the hinge lines. Minor folds E and G verge away from the synclinal axis; (c) vertically plunging fold with a sinistral vergence; (d) vertically plunging fold with a dextral vergence (reproduced with permission from Bell, 1981).

54

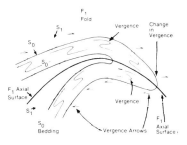

Fig. 3.11 Example illustrating the use of vergence to determine the F_1 hinge and axial plane of a refolded fold.

3.6 Strain in folds

Within a folded layer, a wide variety of strain states can be found (Fig. 3.12). Most folds that develop by buckling of a competent layer have extensional strains on the outer arc and contractional strains on the inner arc. A plane of no strain (*finite neutral surface* — Fig. 3.12a) exists within the folded layer, but this moves down *toward the inner arc* of the fold as it tightens. In the field the extensional strains produce dilatant fractures and veins (see Chapter 7) and cleavages are produced in areas of contractional strains. Flexural slip folds (parallel folds produced by slip of layers over each other, Fig. 3.12b) are characterised by little internal deformation of the layers on the fold limbs, buckling strains (as in Fig. 3.12a) in the hinges, and well-developed slickensides (Fig. 5.4) between competent layers in the fold (Fig. 3.12b). Many folds may have flattening strains superposed upon them (Fig. 3.12c) to produce a near similar fold style. These flattening strains are commonly reflected in the development of well-developed axial-planar cleavages (see Chapter 4).

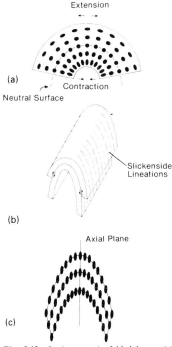

Fig. 3.12 Strain states in folded layers: (*a*) buckle fold produced by bending the layer (i.e. a parallel fold with the thickness perpendicular to the layer remaining constant) and with the development of extensional strains on the outer arc of the fold, a finite neutral surface of no finite strain (note that this will move downwards in the layer as the fold tightens), and contractional strains on the inner arc of the fold; (*b*) flexural slip fold where the folded layers slip past each other (parallel fold). (*c*) a flattened buckle fold which approaches similar style geometry (thickness parallel to the axial plane is constant).

3.7 Folds associated with faults

Many folds are geometrically related to faults. In general, these are passive folds whose geometry is controlled by the fault geometry: these can occur at all scales.

In contractional fault systems, steps in the fault plane require *geometrically necessary folds* to develop in the hanging-wall plate as it moves over the step (Fig. 3.13a). Kink-like and box-fold geometries result. Similarly, in extensional fault systems geometrically necessary folds are generated above changes in fault plane geometry (Fig. 3.13b). Thus, in fault terranes we must expect geometrically necessary folds to be generated wherever there are *changes in dip of the fault plane*. Fault-related folds are further discussed in Chapter 6.

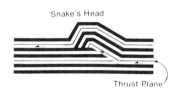

'Snake's Head'

Thrust Plane

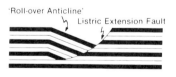

'Roll-over Anticline'

Listric Extension Fault

Fig. 3.13 Geometrically necessary folds generated by changes in dip of fault planes: (*a*) 'snake's head' geometry of kink folded strata over a flat ramp flat thrust fault geometry; (*b*) 'roll-over anticline' fold developed in the hanging-wall above a listric extensional fault.

3.8 Kink bands

Typical examples of kink bands are illustrated in Figs. 3.14 and 3.15. They generally only occur in *strongly foliated anisotropic* rocks (i.e. they are often second or later generation structures developed after a first penetrative cleavage has been formed) and may occur singly or in conjugate pairs. If the latter occurs they may then be used to determine palaeostress orientations (dynamic analysis; see below).

Two forms of kink bands are found:

1 Normal kink bands (Fig. 3.14a) in which there is a volume *decrease* in the kink band.
2 Reverse kink bands (Fig. 3.14b) in which there is a volume *increase* in the kink band.

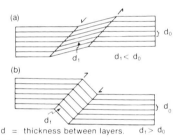

Fig. 3.14 Kink band nomenclature: (*a*) normal kink: note the volume decrease in the kink band; (*b*) reverse kink: note the volume increase in the kink band.

An example of reverse kink bands is shown in Fig. 3.15a. Where the kink bands occupy the total rock volume i.e. at high strains, then the kink axial planes tend to become parallel to the bulk flattening plane (Fig. 3.15b).

Fig. 3.15a Inclined conjugate reverse kink bands in deformed schists. The maximum compressive stress σ_1 is horizontal.

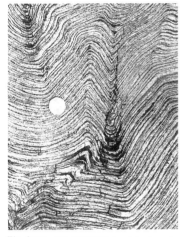

Fig. 3.15b Polyclinal kink bands in strongly deformed psammites. Note the upright kink bands in the central part of the photograph.

Stress analysis using kink bands

Conjugate kink bands may be used to calculate the orientation of principal stresses — (σ_1 maximum principal stress, σ_2 intermediate principal stress and σ_3 least principal stress). σ_1 bisects the *obtuse* angle between kink band boundaries (axial planes). σ_2 is parallel to the line of intersection of conjugate kink band arrays and σ_3 bisects the *acute* angle between kink band boundaries. What to measure in kink bands is outlined in Table 3.3 (p. 58).

3.9 Mapping of folds

When mapping folded rocks it is important to decide what fold structures can be represented on the map and what structures should be represented as minor structures. (In the latter case *detailed sketches and sketch maps* are required to illustrate the significance of minor structures at the outcrop scale: Fig. 2.20a.)

3.9.1 At the outcrop scale

For folds which are found in an individual outcrop or group of outcrops and are too small to plot in detail on your map the following procedure should be adopted:

1 Describe and classify the style of folding (e.g. chevron folding; parallel folding; similar style folding (Figs. 3.3 and 3.4)). Always analyse folds in the profile section looking down the plunge of the fold axis — where possible photographs and sketches should always be made looking down the fold plunge.

Table 3.3 Data to be collected from observations on kink bands.

Structure	What to Measure	What Observations to Record	Results of Analysis
Kink bands (KB) 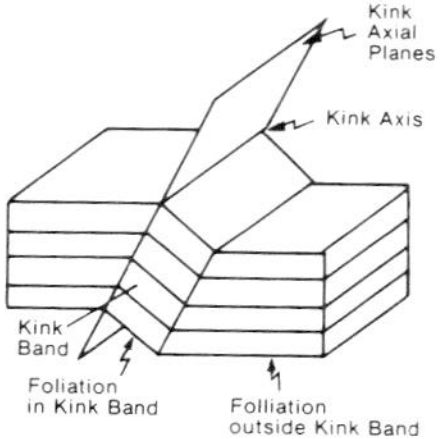	Orientation of foliation outside kink band (Figs. 2.5–2.8).	Nature of foliation that is kinked. Nature of foliation in kink band: Veining (reverse kinks) pressure solution (normal kinks).	Deformation processes during kinking.
	Orientation of kink band axial plane (Figs. 2.5–2.8).	Normal or reverse kink angle between kink band and mean foliation. (Fig. 3.14)	Stress analysis giving σ_1, σ_2, σ_3.
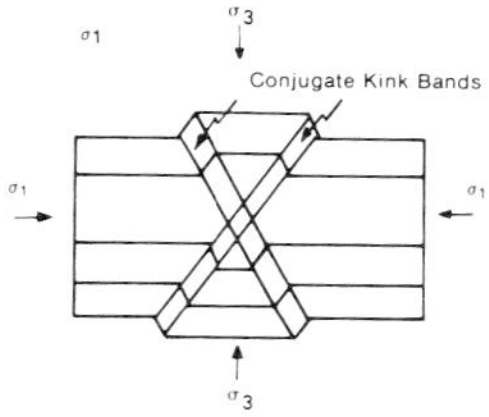	Orientation of conjugate kink band axial plane (Figs. 2.11–2.13).	Angles between conjugate kink bands.	σ_1 bisects obtuse angle between conjugate kink bands.
	Line of intersection of conjugate kink bands (Figs. 2.11–2.13). Fold axes of kink bands (Figs. 2.11–2.13).	Crenulation/kink lineation associated with kink bands.	σ_2 orientation.

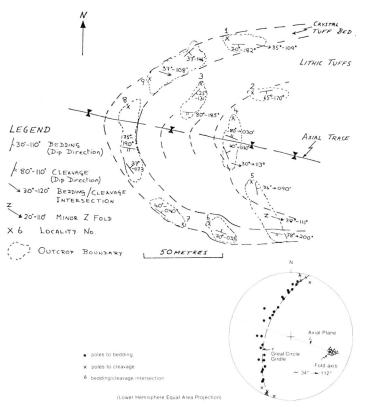

Fig. 3.16   Sketch map of a synclinal fold and a stereographic projection of structural data measured around the fold.

Determine the cylindricity of the fold in order to understand how the folds will project down the plunge of the fold axis (see 9.2.2).

2   Measure the orientations of *both the fold axis and the fold axial surface* (*Figs. 2.5, 2.8c and 2.11*). Measure — the interlimb angle of the folds (Table 3.1) and —

the bedding thicknesses around the fold (this will aid classification — cf. dip isogons Fig. 3.4 and for chevron folds allow simple estimates of shortening across the fold (Ramsay, 1974)).

3   Measure the *orientation of the enveloping surface i.e. sheet dip* (*Fig. 3.1*) — essential in areas

59

of complex small scale folds. Measure bedding around the fold — this will allow better definition of the fold axis by plotting this data on the stereographic projection.

4 Determine the relationships and measure the orientations of cleavages and lineations in the folds (see Chapters 4 and 5 for details).

5 Determine the asymmetry and vergence of the folds (Figs. 3.8, 3.9 and 3.10). Measure the direction of vergence.

6 **Plot on your map** (a) the trend, plunge and asymmetry (vergence) of the minor fold — or mean trend and plunge from a group of minor folds (Table 2.1); (b) the orientation and dip of the fold axial-surface (Table 2.1); (c) the sheet dip of the bedding in the folded outcrop, (d) the orientations of cleavage and lineations developed in the fold.
   Note that your notebook should contain much more structural data than can be plotted at one outcrop on your map.

7 Check that your notebook contains the appropriate sketches and structural data to fully analyse the fold or folds.

3.9.2   On the map

To accurately represent fold structures on your map (e.g. Fig. 3.16) you should endeavour to complete the following steps.

1 Define the fold by mapping sufficient outcrops so that the fold is outlined by (a) bedding form lines, (b) the distribution of lithological units, (c) the distribution of *sheet dip form* lines (cf. Fig. 3.1b). (Note: this is important in areas with abundant small-scale folds.)

2 Attempt to identify the hinge region of the fold — e.g. exposure **8** in Fig. 3.16. Note that in the hinge region of a fold with an axial-planar cleavage, the bedding and cleavage are almost at right angles, whereas on the fold limbs the cleavage is at a lower angle to bedding (see Chapter 4).

3 Map the cleavage/bedding relationships (i.e. cleavage vergence) around the fold.

4 Measure the orientation and asymmetry of small-scale folds around the large fold (Fig. 2.20a). This, together with location of the hinge region and bedding/cleavage relationships, will allow the position of the *trace* of the axial plane to be determined (Fig. 3.16).

5 Measure sufficient structural data from around the fold, to permit the following constructions on a stereographic net (Fig. 3.16):
   (a) Define the great circle of poles to bedding (assuming cylindrical folding). For this a minimum of 15–20 good readings are recommended. This will allow the major fold axis to be defined (Fig. 3.16). (This means that you should always take more readings than can be plotted on the map.) In areas of tight to isoclinal folding, data must be col-

lected from the hinge region in order to define the great circle girdle on the projection.

(b) Measure the cleavage throughout the fold (Fig. 3.16). The mean cleavage plane will approximate to the fold axial plane. (A minimum of 20 readings are recommended for this.)

(c) Measure the bedding/cleavage intersection and minor folds around the fold hinge (Fig. 3.16). These should be approximately parallel to the major fold axis (a minimum of 20 readings are recommended).

6 Identify and classify the style of folding in the map area (e.g. parallel folding generated by flexural slip; chevron folding; or similar style folding—Figs. 3.3 and 3.4), particularly by observing minor and mesoscale folds in outcrop. The outcrop pattern on your map and your cross-section (see 9.2) must reflect the style of folding.

Representation of a large fold on your map should always be supplemented by a sketch cross-section showing structural relationships.

Table 3.4 summarises the data to be collected when mapping folds.

Table 3.4 Data to be collected from observations when mapping folds from a single phase of deformation.

Structure	What to Measure	What Observations to Record	Results of Analysis
Fold Axial Plane / Fold Axis	Orientation of fold axial surface (dip direction) (Figs. 2.5–2.8).	Nature of axial surface. Relationships of axial planes in a group of folds.	Orientation of fold structure. (Table 3.2)
	Orientation of fold axis (plunge) (Figs. 2.11–2.13).	Nature of hinge line —straight or curved. Relationships of hinge lines in a group of folds.	
vergance / Z asymetry	Vergence (azimuth) (Figs. 3.9 & 3.10).	Vergence and sense of asymmetry. S, Z, M (parasitic folds) facing. (Figs. 3.8, 3.5, 3.10)	Vergence boundaries. (Fig. 3.11) Axes of major fold structures. (Fig. 3.8) Tectonic transport direction. (Fig. 3.4)
Similar Fold / Dip Isogens	Profile section of fold (Figs. 3.2–3.4).	Thickness changes in profile section. Cylindricity. Fold type. (Figs. 3.3, 3.4, 3.5)	Fold classification: 2D or 3D, dip isogons. (Fig. 3.4) Projection of fold down plunge. (Fig. 9.6)
Cleavage / S_1 / S_0	Cleavage orientations around the fold (Fig. 4.3).	Nature of cleavage. (Fig. 4.1)	Mean cleavage approximates to fold axial plane. (Fig. 4.3b) Deformation mechanisms.

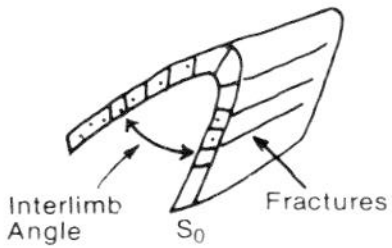	Fracture patterns around fold (Figs. 7.4 & 7.5).	Nature of fractures — veining. (Figs 7.1 and 7.6)	Deformation mechanisms.
	Interlimb angle. (Table 3.1)	Nature of limbs — planar — curved. (Fig. 3.3)	Shortening across fold, (chevron folds).
	Limb lengths. Strain of deformed objects around the folded layer(s) (Fig. 3.12).	Asymmetry. (Fig. 3.8) Nature of strain in deformed objects. (Appendix III)	Quantification of asymmetry. Strain distribution, mechanisms of folding. (Fig. 3.12)

4
Foliations

Foliation is a planar rock fabric. In this Handbook we are concerned with tectonic foliations that are usually produced by deformation and recrystallisation of the mineral grains within the rock to produce a preferred orientation (as opposed to bedding-parallel fabrics induced by compaction during burial). Most foliations (with the exception of fracture cleavages) are penetrative on a mesoscopic scale, i.e. they penetrate the whole rock, unlike joints or fractures, which have little or no effect on the rock mass away from the fracture zone.

Planar surfaces in a rock mass are designated 'S' surfaces (excluding joints and fractures). Bedding is 'S$_0$'; the first cleavage is 'S$_1$'; the second cleavage is 'S$_2$' and so on (see mapping symbols, Section 2.6). The subscripts indicate the relative chronology of the surface. Rocks with strong planar tectonic fabrics are termed '*S tectonites*'.

4.1   Common foliations

The type of tectonic foliation developed in a rock will be strongly dependent upon the conditions of deformation (temperature, confining pressure, differential stress and strain rate) and upon its composition. Rocks which have abundant platy minerals (clays and micas) will tend to develop penetrative foliations that give a strong directional parting or *fissility* to the rock, whereas non-platy monomineralic rocks such as limestones or quartzites will tend to develop spaced cleavages (visible discrete foliation planes) or grain shape fabrics (see Fry, 1984). Listed below are the more common types of foliations found in rocks.

In non-metamorphosed to low-grade metamorphic rocks:

1   *Slaty cleavage:* Penetrative foliation occurring in fine-grained incompetent units e.g. mud rocks and imparting a strong *fissility* to the rock — in the hand specimen no visible minerals or segregations of minerals to be seen in the plane of cleavage (Fig. 4.1a).

2   *Crenulation cleavage:* Foliation produced by microfolding (crenulation folding) *of a pre-existing foliation* commonly associated with a segregation of minerals which can be seen as bands in the plane of cleavage (Fig. 4.1b). May be penetrative

Fig. 4.1a Penetrative slaty cleavage in mudstones showing a well-defined fissility.

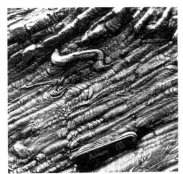

Fig. 4.1b Crenulation cleavage (parallel to penknife) formed by microfolding of thinly interbedded psammites (sandy units) and pelites (originally muddy units). The limbs of the crenulation microfolds in the psammitic layers are disrupted and show partial dissolution features (pressure solution) that result in a preferential concentration of phyllosilicates in bands that define the crenulation cleavage planes.

Fig. 4.1c Well-developed fracture cleavage consisting of a closely-spaced array of vertical fractures in sandstones.

Fig. 4.1d Pressure-solution cleavage showing disruption of the light-coloured psammitic layers by dark pressure-solution seams.

in fine-grained rocks. Common in slates, phyllites and schists (see Fry 1984).

3   *Fracture cleavage:* A non-penetrative foliation consisting of persistent, closely-spaced fractures (Fig. 4.1c). May occur in sandstones, massive limestones and in igneous rocks.

4   *Pressure-solution cleavage:* A *spaced cleavage* which produces a mineral segregation (often associated with microfolding) and dark seams of insoluble material that impart a prominent striping to the rock (Fig. 4.1d).

In higher grade metamorphic rocks:

5 *Schistosity:* A penetrative/non-penetrative foliation with visible phyllosilicates and mineral segregation into bands parallel with the foliation (Fig. 4.2a). (Note that schistosity is commonly parallel with bedding.) Schistosity is often folded by later crenulation cleavages.

6 *Gneissic foliation:* A foliation in coarse-grained rocks, consisting of impersistent laminae and segregations of mineral grains (Fig. 4.2b). (Note that the gneissic foliation is often parallel–sub-parallel with the lithological banding.)

7 *Mylonitic foliation:* A penetrative foliation developed in zones of high shear strain such as faults and shear zones. It is characterised by tectonic reduction in grain size (Fig. 4.2c) — often resulting in extremely fine-grained, almost slaty rocks.

Fig. 4.2b Gneissic foliation showing the crude banding of quartzo-feldspathic and of mafic segregations.

Fig. 4.2a Schistosity showing segregation into phyllosilicate layers and quartzo-feldspathic layers.

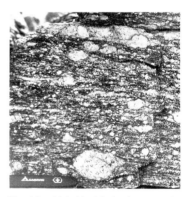

Fig. 4.2c Mylonitic foliation in a strongly sheared granite showing a well-developed planar fabric with deformed feldspar porphyroclasts in a matrix of fine-grained streaked out quartz and feldspar.

66

4.2 Axial-planar foliations

In most cases foliations are approximately axial planar to the folds of the deformation phase which produced the foliation. This general relationship is shown in Figs. 4.3a and b. The foliation plane approximates to the plane of finite flattening (XY plane of the strain ellipsoid) for the deformation that produced the foliation (Fig. 4.3a). This is a general rule that can be applied to folded sequences, but it breaks down within shear zones where the foliation plane is not parallel to the finite flattening plane outside the shear zone.

Fig. 4.3a Axial-planar cleavage in folded silt-stones.

4.2.1 Fanning and refracted foliations

In most cases the cleavage, e.g. slaty cleavage, is not strictly axial planar to the fold, but in fact, fans around the fold (Fig. 4.3b). This arises from the difference in competence or stiffness of the beds that are undergoing folding. In the case of interbedded pelites (fine-grained mud rocks) and psammites (coarser-grained sandy rocks) (Fig. 4.4a), on the fold limbs the cleavage is at a low angle to bedding in the slaty lithologies but refracts to a high angle to bedding in the sandstones. This zig-zag cleavage refraction reveals important differences in lithological compositions, e.g. the cleavage plane curves toward parallelism with the top of a graded bedding unit, and as such, can be used to determine way-up (Fig. 4.4a).

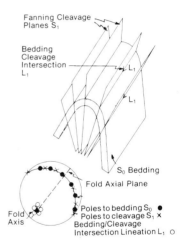

Fig. 4.3b Fanning axial-planar foliation around a folded layer. The intersection lineation L_1 is parallel to the major fold axis. The stereographic projection shows the plot of poles to bedding (solid circles), poles to cleavage planes (crosses) and L_1 lineations (open circles). The major fold axis (triangle) is defined by the pole to the bedding great circle.

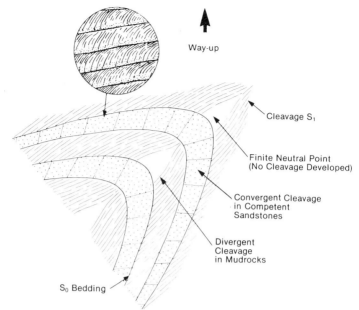

Way-up

Cleavage S₁

Finite Neutral Point
(No Cleavage Developed)

Convergent Cleavage
in Competent
Sandstones

Divergent
Cleavage
in Mudrocks

S₀ Bedding

Fig. 4.4a Refraction of cleavage around folded layers of different competencies. Note the characteristic flattening of the cleavage in the more pelitic layers and the higher angle between cleavage and bedding in the more competent sandy layers. In normally graded beds this distinctive pattern of cleavage flattening upwards may be used to determine way-up.

An example of cleavage refraction is shown in Fig. 4.4b.

Fanning cleavage is designated as occurring in either convergent or divergent cleavage fans (Fig. 4.4a). In strongly contrasting lithologies, e.g. slate and sandstone (Fig. 4.4a), '*anomalous*' cleavage orientations are found in the hinge region of folds; in particular, an area of no cleavage occurs at the *finite neutral point* (Fig. 4.4a). This cleavage pattern is found in the less competent units, e.g. slates, and arises from the finite strain patterns around the fold hinge zone.

Fig. 4.4b Cleavage refraction in interbedded sandstones and shales.

68

4.3 Foliations and folds

The relationship between foliation and folds is extremely useful in determining the presence and location of major fold structures. Fig. 4.5 shows how the antiform hinge is located. Note that in the overturned limb the bedding is steeper than cleavage and vice versa in the non-inverted limb. There is a near orthogonal relationship between bedding and cleavage in the fold hinge region. Cleavage bedding relationships tell you the *structural* way-up. Younging evidence (e.g. Figs. 1.1 and 4.4a) from sedimentary structures is needed to tell the *stratigraphic* way-up.

The intersection of the foliation surface (S_1) with the bedding plane (S_0) produces a lineation (L_1) parallel with the fold axis (Fig. 4.3b and 3.16). It is essential that this lineation is systematically recorded during a mapping programme because it is a measure of the plunge of the folds (see Table 3.4).

4.3.1 Transposition

Transposition is the rotation of a pre-existing foliation or bedding into parallelism or near parallelism with the fold axial plane. This produces a new transposed layering (see Hobbs *et al.*, 1976), which may also incorporate mineral segregation and redis-

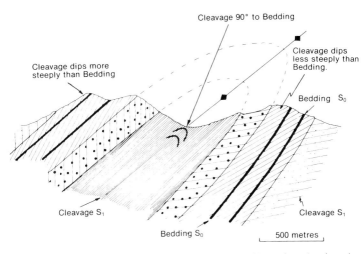

Fig. 4.5 An illustration to the use of bedding/cleavage relationships to determine the major structure and to locate the major fold hinge.

tribution and hence result in a new metamorphic layering (Fig. 4.6a). On a small scale the orientation of this new layering often does not represent the gross orientation of the larger lithological units. An example of transposition in greenschist-facies rocks is shown in Fig. 4.6b.

Fig. 4.6b Example of transposition. Thin sandy beds are transposed into the foliation (vertical). The original bedding planes (S_0) cannot be traced across the photograph.

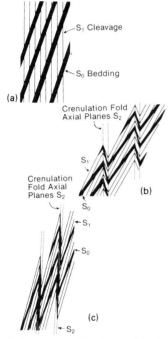

Fig. 4.6a Transposition of a pre-existing foliation S_1 and bedding S_0 by crenulation folding to produce a transposed foliation S_2. (a) Bedding S_0 and first cleavage S_1; (b) Asymmetric crenulation folding with rotation of both S_1 and S_0 into the fold limbs; (c) Increased rotation of S_1 and S_0 towards parallelism with the S_2 fabric (crenulation fold axial planes). Pressure solution on the fold limbs, recrystallisation and metamorphic growth will enhance the S_2 fabric.

4.3.2 Superposed foliations

In an area that has been affected by two deformations, the first foliation (S_1) becomes folded and a new foliation (S_2) is developed (Fig. 4.7). The second foliation is usually a *crenulation cleavage* which intersects the first foliation to produce a *crenulation lineation* L_2 in S_1. The crenulation foliation may also intersect the bedding (S_0) to produce a crenulation lineation L_0^2 in S_0 (see Table 8.1). In areas of only two phases of folding, the crenulation foliation S_2 (approximately axial planar to the second phase of folds) will be fairly constant in orientation, whereas the orientation of L_2 will vary according to the orientation of S_1 within the first-phase fold structures (e.g. fanning S_1 foliation). The orientation of the crenulation cleavage/bedding inter-section L_0^2 will vary significantly *according to the orientation of the bedding S_0 in the first-phase fold structures.*

Transposition (Section 4.3.1) is a common feature of microfolding associated with crenulation-type cleavages produced by a second deformation.

4.4 Mapping foliations

Foliations should be systematically mapped and plotted on your field slips in the same way as bedding. Measurement techniques are the same as those employed for any 'S' surface (Section 2.3.2). Data to be collected if one tectonic cleavage or foliation are developed are listed in Table 4.1. The importance of meas-uring the orientation of *cleavage* S_1 and the *bedding/cleavage intersection* L_1 cannot be overemphasised. If two tectonic cleavages are present then Table 4.2 outlines the additional data to be collected.

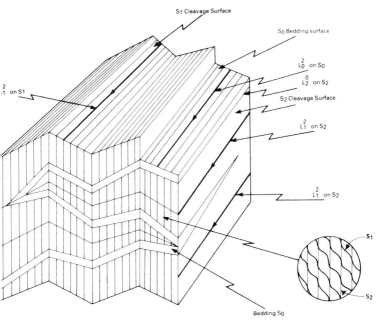

Fig. 4.7 Superposed foliations. A second cleavage—a crenulation cleavage S_2 is developed by microfolding of the first cleavage S_1 (a slaty cleavage).

71

Table 4.1 Data to be collected from observations on the first cleavage (or schistocity), S_1 (commonly a slaty cleavage).

Structure	What to Measure	What Observations to Record	Results of Analysis
	S_1 dip direction (or strike and dip) (Figs. 2.5–2.7). Of cleavage or schistosity	Orientation of cleavage relative to bedding. (Figs 2.15 and 4.5) Sense of vergence. (Figs 3.9, 3.10) Facing. Cleavage refraction. (Fig. 4.4) Nature of cleavage. (Fig. 4.1)	Position relative to fold axis. (Fig. 4.5) Vergence of structure. (Fig. 3.9, 3.10) Facing of structure. Mean cleavage approximates to fold axial plane.
	L_1 bedding lineation on cleavage plane (plunge) (Figs. 2.11–2.13).	Nature of lineation. (Figs 5.1 to 5.4)	Orientation of fold axis (b_1 axis). (Fig. 4.3b)
	Mineral stretching ML_1 lineation on cleavage plane (plunge) (Figs. 2.11–2.13).		Orientation of stretching axis $\approx$ X axis of bulk strain ellipsoid (a_1 axis). (Fig. 3.2 and Appendix III)
	Orientation and magnitude of strain of deformed objects in the cleavage plane (Appendix A.III).	Nature of strain relative to cleavage. (Appendix III)	XY plane of strain ellipsoid. (Appendix III)
In polyphase terranes.	L_2 on S_1. The intersection of subsequent cleavages on the first cleavage plane, i.e. crenulation lineations (plunges) (Figs. 2.11–2.13).	Nature of intersection of second-phase cleavage with first cleavage.	Orientation of second-phase fold axes (for folded first-phase cleavage planes). (Fig. 8.3)

Table 4.2 Data to be collected from observations on the second cleavage S_2 (commonly a crenulation cleavage or schistosity).

Structure	What to Measure	What Observations to Record	Results of Analysis
S_2 (Commonly a crenulation cleavage or Schistosity). Crenulated S_1. L_2 intersection of S_1 on S_2	Dip direction (or strike and dip) (Figs. 2.5–2.7) of S_2.	Nature of S_2 cleavage: orientation of S_2 cleavage relative to S_1 cleavage and relative to bedding S_0. (Fig. 4.7) Sense of vergence. (Fig. 3.9, 3.10) Facing on cleavage.	Position relative to F_2 fold axis. (Fig. 4.5) Mean cleavage approximates to F_2 axial plane. Vergence and facing of F_2 structure.
	L_2 Intersection of first cleavage on second cleavage plane S_2 (Figs. 2.11–2.13).		Orientation of F_2 fold axis (b_2 axis) of folded S_1 surface. (Figs. 4.7, 5.1d)
	L_2^1 Intersection of bedding on second cleavage plane (Figs. 2.11–2.13).		Orientation of F_2 fold axis for folded bedding S_0 surface (note that this depends upon bedding S_0 and F_1 limbs). (Fig. 8.3)
	Mineral stretching ML_2 lineation on cleavage plane (Figs. 2.11–2.13).	Nature of lineation. (Fig. 5.1 to 5.4)	Orientation of stretching axis $\approx$ X axis of bulk strain ellipsoid for F_2 deformation (a_2 axis). Fig. 3.2 and Appendix III
	Orientation and magnitude of strain in deformed objects in the cleavage plane (Appendix III).	Nature of strain relative to cleavage. (Appendix III)	XY plane of F_2 strain ellipsoid. (Appendix III)

5
Linear structures

A lineation is a linear rock fabric that may result from the intersection of two planar features, from the alignment of mineral grains, crystals or clasts within the rock, from linear shape fabrics of grains and clasts, or from the parallel alignment of tectonic elements such as minor fold or crenulation axes or slickenside-groove features. Here we are concerned with tectonic lineations — primary or depositional lineations are discussed by Tucker (1982). Lineations include bedding/cleavage intersections, crenulation lineations, minor fold axes, mineral stretching lineations, slickensides, grooving on fault and fracture planes, and boudin axes. Rocks with a penetrative linear fabric are termed *L tectonites*. Lineations are commonly used in complex deformed areas to define *sub-areas of structural homogeneity*. *Reorientation* of earlier linear structures can be an important indication of later deformation.

5.1 Lineations associated with folding

5.1.1 *Bedding cleavage intersection*

The most distinctive form of linear structure in simply folded areas is the bedding(S_0)/cleavage(S_1) intersection (L_1) (Fig. 5.1a). This basic geometry is illustrated in Fig. 5.1b. The intersection of bedding surfaces and cleavage is parallel or approximately parallel to the fold axis b_1. This lineation may be measured directly on the outcrop (Section 2.3.3, Figs. 2.11–2.12) or determined on the stereonet from bedding and cleavage measurements. Caution: the cleavage may not be precisely axial planar to the fold and hence the bedding/cleavage intersection L_1 may not always be exactly parallel to the major fold axis b_1. A fold with an oblique cleavage (possibly as much as 20° from the axial plane) is known as a transected fold. Careful observation of cleavage relationships in minor folds will indicate whether the folds are transected or not.

5.1.2 *Crenulation lineations*

Crenulation lineations are formed by the intersection of a crenulation cleavage (e.g. S_2) with an earlier foliation (e.g. S_1) (Fig. 5.1c). A crenulation cleavage *requires a pre-existing foliation* in order to develop – this may be a bedding fabric, a slaty cleavage or an earlier schistosity (Fig.

Fig. 5.1a Bedding/cleavage intersection lineation L_1 as observed on the bedding plane in deformed mudstones.

5.1d). Commonly, the presence of a crenulation lineation on the slaty cleavage is a good indication of a second deformation. In some situations a crenulation cleavage may also result from progressive deformation during a single deformation.

Fig. 5.1c Crenulation lineation developed on the schistosity surface (or S_1 surface) of crenulated psammitic schists.

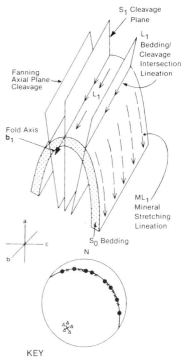

KEY

• poles to bedding

+ fold axis

△ L_1 bedding/cleavage intersection lineation

× poles to cleavage

Fig. 5.1b Folded layer S_0 with a fanning axial-planar cleavage S_1 showing the development of the bedding/cleavage intersection lineation L_1 which is parallel to fold axis or hinge line. A mineral stretching lineation ML_1 is developed on the fold limbs and at 90° to the fold axis. The stereographic projection shows the plot of poles to bedding (solid circles), bedding/cleavage intersections L_1 (triangles) and poles to fanning cleavage planes (small crosses). Fold axis is defined by the pole to the bedding great circle.

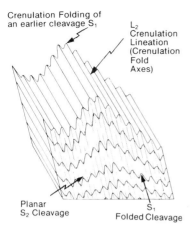

Crenulation Folding of an earlier cleavage S_1

L_2 Crenulation Lineation (Crenulation Fold Axes)

Planar S_2 Cleavage

S_1 Folded Cleavage

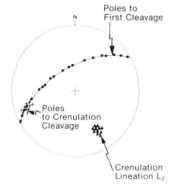

Poles to First Cleavage

N

Poles to Crenulation Cleavage

Crenulation Lineation L_2

Fig. 5.1d Crenulation folding of a cleavage S_1, showing the development of a crenulation cleavage S_2 and an intersection (S_2 on S_1) lineation L_2. The stereographic projection shows poles to the first cleavage S_1 (solid circles), poles to second cleavage S_2 (crosses) and the crenulation lineation L_2 (triangles). The major F_2 fold axis is defined by the pole to the S_1 cleavage great circle.

For second-phase deformations, the crenulation cleavage(S_2) / slaty cleavage(S_1) intersection lineations

(L_2) may be analysed in a similar fashion to bedding/cleavage intersections (Fig. 5.1d).

5.1.3 Pencil cleavage lineations

In some rocks (often mudstones and siltstones) there is a strong bedding plane (S_0) fabric which, when folded and cut by a slaty cleavage causes the rock to break into pencil-like slivers (Fig. 5.2). This is known as a pencil cleavage lineation and should be measured and analysed in the same way as the normal bedding/cleavage intersection (Section 5.1.1).

Fig. 5.2 Pencil cleavage lineation in a dolomitic siltstone as a result of the intersection of a strong bedding-parallel fabric and a strong tectonic cleavage.

5.2 Mineral stretching and elongation lineations

5.2.1 Mineral stretching lineations

Mineral stretching lineations (ML_1) are formed by elongate mineral grains

Fig. 5.3 A mineral stretching lineation ML_1 in a sheared granite showing elongate quartz grains and feldspar aggregates.

Fig. 5.4 Fibrous quartz slickenside lineations developed on the limb of a flexural slip fold.

(Fig. 5.3) and *elongation lineations* are formed by alignment of deformed pebbles and fossils. Both of these types of lineation can be either parallel to the fold axis (*b lineations*) or at high angles to it (*a lineations*) (see Fig. 5.1b). Careful observation of the relationships between the *minor fold axes and linear structures* is essential.

5.2.2 Mineral lineations associated with flexural slip folding

In flexural slip folds (Fig. 3.12b) internal slip of layers over one another produces slickensides, grooving and/or mineral stretching lineations, all of which are appproximately 90° to the fold axis. These lineations indicate the direction and sense of slip between the layers. The slickenside lineations are commonly fibrous accretion steps (Fig. 5.4).

5.3 Lineations formed by boudins, mullions or rodding

5.3.1 Boudins

In folds whose limbs are strongly stretched, the more competent units will tend to neck into elongate lozenges called *boudins* (Fig. 5.5a). Similarly, if there is strong flattening perpendicular to a cleavage or schistosity plane, then competent units will neck into boudins (Fig. 5.5b) The boudin axis will tend to be parallel to the '*b*' tectonic axis (i.e. the fold axis).

5.3.2 Mullions

Mullions form in a similar fashion to boudins in that they are usually parallel to the '*b*' tectonic axis and occur when the *interface* between incompetent and competent material is deformed (Fig. 5.5a).

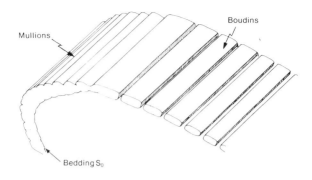

Fig. 5.5a Lineations formed by boudinage and by mullions. In simple boudin structures the lineations tend to parallel the major fold axes.

Fig. 5.5b Boudin necks of amphibolite in quartzo-feldspathic gneisses.

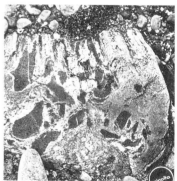

Fig. 5.5c L tectonite with rods of amphibolite in a quartzo-feldspathic gneiss.

5.3.3 Rodding

Rods are stretched and elongated fragments of competent material in a ductile matrix. Pebbles, quartz veins or dyke fragments commonly form rods that are parallel to either the '*a*' or '*b*' tectonic axis. Rodding produces L tectonites (Fig. 5.5c). Careful examination of the relationships between rodding and other structures such as folds is needed to determine whether the rods are parallel to the '*a*' or '*b*' direction.

5.4 Lineations associated with faults

Slickensides and grooving are commonly associated with brittle faulting. Slickensides are often composed of fibrous crystals that stretch from one side of the fault plane to the other (Fig. 5.6a). In carbonates, linear structures called slickolites are formed by pressure solution of the bumps (asperities) in the fault plane and reprecipitation of fibrous calcite in the gaps (Fig. 5.6c). Grooves are formed by comminution (grinding down) and solution as the two fault surfaces slide over one another (Fig. 5.6d).

(a)

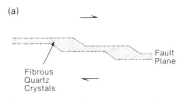

Fig. 5.6a Slickensides developed as fibrous crystals of quartz joining opposite sides of a stepped fault plane.

Fig. 5.6b Slickensides of fibrous quartz in sandstones. The movement direction is up in the photograph.

(c)

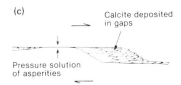

Fig. 5.6c Slickolites developed in limestones by pressure solution which removes the bumps (asperities) in the fault surface, and redeposits the calcite in the spaces between the stepped fault surfaces.

Fig. 5.6d Grooving on a fault plane in limestones.

79

5.5 Mapping linear structures

Linear structures are extremely important in structural mapping as they can be used to separate deformation phases and to determine the kinematics of deformation. In polyphase deformed terranes the consistency of particular intersection lineation orientations is a key factor in the subdivision of a map into structurally homogeneous sub-areas (see Section 8.3). They must be systematically measured and recorded whilst mapping.

Note:

1 The orientations of linear structures are measured as plunges (or less commonly as pitches, Section 2.3.3). They plot as points on the stereographic projection (Fig. 5.1b).
2 Intersection lineations e.g. L_1 (S_1/S_0) (parallel to b) must be distinguished from mineral stretching or elongation lineations e.g. ML (generally parallel to a).

Data to be collected during the mapping of lineations are outlined in Tables 5.1, 5.2 and 5.3.

Table 5.1 Data to be collected from observations on intersection lineations L_1, L_2, etc.

Structure	What to Measure	What Observations to Record	Results of Analysis
L_1 Bedding S_0/cleavage S_1 intersection on either S_0 or S_1 surfaces. 	Plunge of lineation L_1 (note: orientation data for S_0 and S_1 also required) (Figs. 2.11–2.13). Strain of deformed objects parallel to lineation (Appendix III).	Nature of lineation. (Figs. 5.1 to 5.4) Orientation and nature of bedding and cleavage. (Fig. 4.1) Nature of strain. Fibre growth parallel to lineation. Fractures parallel to lineation.	Lineation generally parallels F_1 fold axis (b_1). (Fig. 5.1b) Y axis of F_1 strain ellipsoid. (Appendix III)
L_2 First cleavage S_1/second cleavage S_2 intersection on either S_1 or S_2 surfaces (S_2 is generally a crenulation cleavage). 	Plunge of lineation L_2 (note data for S_0, S_1 and S_2 are also required) (Figs. 2.11–2.13). Strain of deformed objects parallel to lineation (Appendix III).	Nature of lineation. (Figs. 5.1 to 5.4) Orientation and nature of S_0, S_1, and S_2. (Fig. 4.1) Nature of strain. Fibre growth parallel to lineation. Fractures parallel to lineation.	Lineation generally parallels F_2 fold axis b_2 (for S_1 surfaces). (Fig. 5.1b) Y axis of F_2 strain ellipsoid. (Appendix III)

Table 5.2 Data to be collected from observations on mineral stretching lineations ML_1, ML_2.

Structure	What to Measure	What Observations to Record	Results of Analysis
ML_1 Mineral stretching lineation in S_1. 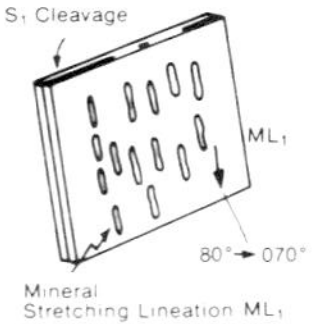	Plunge of ML_1 lineation (orientation data for S_0 and S_1 also required) (Figs. 2.11–2.13).	Nature of lineation (nature of bedding and cleavage also required). Overgrowths parallel to lineation, fibre directions.	Lineation generally parallels the X axis of the F_1 finite strain ellipsoid, ('a_1' tectonic axis). (Fig. 3.2; Appendix III)
	Strain of deformed objects parallel to lineation (Appendix III).	Nature of strain. Fibre overgrowths.	X axis of F_1 finite strain ellipsoid.(Appendix III)
ML_2 Mineral stretching lineation in S_2. 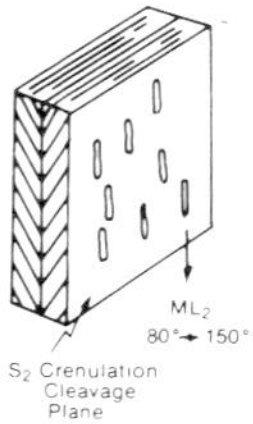	Plunge of lineation (orientation data for S_0, S_1, and S_2 are also required) (Figs. 2.11–2.13).	Nature of lineation (nature of S_1, also required). Overgrowths parallel to lineation, fibre directions.	Lineation generally parallels X axis of the F_2 finite strain ellipsoid (a_2 tectonic axis), (fig. 3.2; Appendix III)
	Strain of deformed objects parallel to lineation (Appendix III).	Nature of strain. Fibre overgrowths.	X-axis of F_2 finite strain ellipsoid. (Appendix III)

Table 5.3 Data to be collected from observations on lineations associated with faults.

Structure	What to Measure	What Observations to Record	Results of Analysis
Grooving (no crystal fibre growth).	Plunge of lineation. Orientation of fault surface. Orientation of displaced units (Figs. 2.11–2.13).	Nature of grooving. Fault rocks. Sense of movement from steps in fault plane. Width of fault zone. Displacement. Stratigraphic separation.	Sense and direction of movement of fault (solutions for exact displacements are not common).
Slickensides (crystal fibre growth).	Plunge of lineation. Orientation of fault surface. Orientation of displaced units (Figs. 2.11–2.13).	Nature of fibre growth. Sense of movement from fibres and steps in fault plane. Fault rocks. Width of fault zone. Displacement. Stratigraphic separation.	Sense and direction of movement of fault (solutions for exact displacements are not common).
Slickolites (Fig. 5.6c)	Plunge of lineation. Orientation of fault surface. Orientation of displaced units. Movement direction 90° to accretion steps.	Nature of fibre growth. Sense of movement from fibres and steps in fault plane. Fault rocks. Width of fault zone. Displacement. Stratigraphic separation.	Sense and direction of movement of fault (solutions for exact displacements are not common).

6
Faults and shear zones

Faults

Brittle to semi-brittle faults are planar discontinuities along which significant displacement has occurred. They generally form in the upper 10–15 km of the crust.

6.1 Classification and description of faults

Numerous classification schemes for faults have been erected based on the *dip of the fault plane and the direction of slip*. In many cases, it is not possible to find the exact displacement as this requires knowledge of the location of matching points on either side of the fault plane. It is not easy to determine the direction of slip if the fault plane is not exposed. In this Handbook two schemes are referred to: (*a*) *Anderson's Dynamic Classification* which relates to the stress systems responsible for the faulting and (*b*) a simple descriptive scheme based upon the geometry and separation across a fault plane.

6.1.1 *Anderson's dynamic classification of faulting*

Anderson's (1951) dynamic classification of faults (Fig. 6.1) is based

on the fact that no shearing stress can exist at the Earth's surface, hence, for faulting to occur close to the Earth's surface, one of the principal stresses (σ_1 σ_2 or σ_3) must be perpendicular to the Earth's surface, and therefore vertical.

Normal faults σ_1 is vertical and σ_2 and σ_3 are horizontal. The dips of the fault planes are $\sim 60°$.

Wrench or strike-slip faults σ_2 is vertical and σ_1 and σ_3 are horizontal. In this case the fault planes are vertical and the movement direction is horizontal, i.e. strike-slip.

Reverse faults σ_3 is vertical and σ_1 and σ_2 are horizontal. The fault planes dip at approximately 30° to the horizontal.

Note that the angle bisected by σ_1 between conjugate fault planes is a function of the material properties of the rocks undergoing faulting and can vary between 45° and 90°. 60° is taken as a typical value.

6.1.2 *Geometric classification and description of faults*

This classification is based upon the sense of movement (separation)

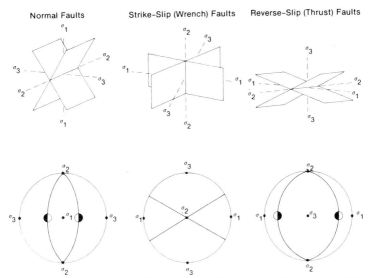

Fig. 6.1 Anderson's dynamic classification of faults with the stereographic projections of the faults and stress systems. The solid half circles indicate the side of the fault block that has moved down.

and direction of slip across the fault plane and is as follows:

1 *Extension* faults—e.g. normal faults (Fig. 6.2a).
2 *Contraction* faults—e.g. reverse faults, thrust faults (Fig. 6.2b).
3 *Strike-slip*—e.g. *wrench* faults, transform faults (Fig. 6.2c and d).

Listric faults Many fault planes are curved, not planar, at depth. Fault planes which are *concave upwards,* and have a fault plane which flattens out at depth are termed *listric* faults. Two types are found: the *listric extension fault* (Fig. 6.3a) is a curved fault which may be divided into a high-angle extension fault, medium-angle

extension fault and bedding plane or sole fault segments. With the high-angle and medium-angle fault sections, stratigraphy is omitted and younger rocks overlie older rocks. A listric extension fault is shown in Fig. 6.3b.

A listric contraction fault (Fig. 6.3c) is a curved fault in which the steep, often sub-vertical segment is a high-angle contraction fault. In both the steep and middle segments, *older rocks* overlie *younger rocks,* whereas there is little or no repetition by the sole fault. More complicated geometries arise in systems where faults may have several listric sections and are *linked* to other faults (see 6.3 and 6.4).

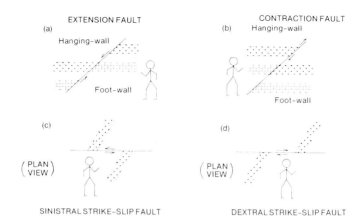

Fig. 6.2 Geometric classification of faults based upon separation across the faults *and with a knowledge of the slip direction on the fault (i.e. from slickensides)*. For an observer standing on the fault plane, the *hanging-wall* is the rock unit above the fault plane and the *foot-wall* is the rock unit below the fault plane; (*a*) a dip slip extension fault as seen in a cross-section producing stratigraphic omission. As viewed by the observer, the hanging-wall fault block has moved down relative to the foot-wall fault block; (*b*) dip slip contraction fault as seen in a cross-section producing stratigraphic overlap. As viewed by the observer, the hanging-wall has moved up relative to the foot-wall; (*c*) sinistral strike-slip or wrench fault in plan view. As seen by the observer standing on the nearest fault block, the fault block on the other side of the fault plane has moved to the left; (*d*) dextral strike-slip or wrench fault in plan view. As seen by the observer standing on the nearest fault block, the fault block on the opposite side of the fault plane has moved to the right.

6.2 Fault displacements

In many cases *exact* fault displacements cannot be determined in the field, but for most faults the following data can be collected:

1 *Direction of movement:* this can be determined from grooving, slickensides, stretched crystal fibres and slickolites on the fault plane (Fig. 6.4). Movement lineations should be plotted as ancillary symbols on the fault plane symbol (see Table 2.1, p. 42).

2 *Sense of movement:* is determined from stratigraphic relationships (e.g. older rocks over younger rocks), from the *apparent* offsets of marker units, dykes and other faults. Great care, however, needs to be exercised, when faults cut *already folded* strata. The sense of movement should be plotted on the stereographic projection, using a partially filled circle with the solid segment on the downthrown side e.g. Fig. 6.1. Fig. 6.5 shows an example of a contraction fault where sense of displacement is from left to right.

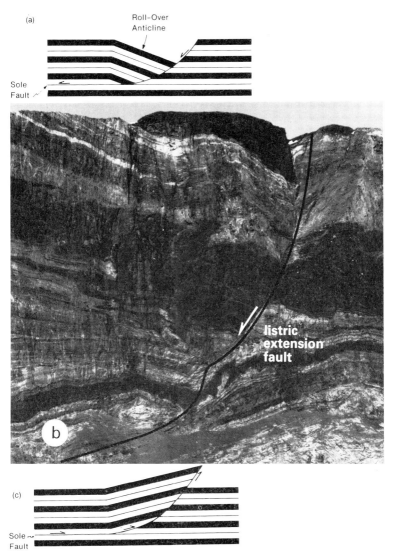

Fig. 6.3 Listric faults — fault planes concave upwards: (*a*) listric extension fault. (*b*) listric extension fault in marbles and pelites, West Greenland; field of view *ca*. 300 metres. (*c*) listric contraction fault.

Fig. 6.4 Fibrous quartz slickensides developed on a fault plane.

Fig. 6.5 Contraction fault in sandstones and shales showing stratigraphic overlap and contraction from left to right. Field of view *ca* 75 m.

3 *Stratigraphic separations:* It is usually possible to either measure or estimate the stratigraphic separation across a fault, using measured or estimated stratigraphic thicknesses of strata affected by the fault: e.g. stratigraphic extension across an extensional fault and stratigraphic contraction across a contractional fault (Fig. 6.2). Where possible the stratigraphic separation should be

marked adjacent to the fault symbol on the map (see Table 2.1).

4 *Rotation* is usually difficult to assess in the field and requires a knowledge of displaced points on both sides of the fault plane.

Once the movement direction of a fault has been determined, the simple classification of faults can be refined to specify the direction of slip as illustrated in Fig. 6.6. Fig. 6.6 also illustrates the terms *throw and heave*.

6.3 Extension faults

The term *extension fault* is preferred to the more commonly used *normal fault* because it refers to the effect of the fault (i.e. it extends the strata). An example of conjugate extension faults is shown in Fig. 6.7. Extension faults may be planar (Fig. 6.2) or listric (Fig. 6.3).

6.3.1 Extensional fault systems

Extensional faults may occur in linked systems, of which two main types are found:
(a) domino fault systems (Fig. 6.8a) of planar rotational extension faults linked by a basal detachment;
(b) listric extensional faults (Fig. 6.8b) producing rotation of the hanging-wall blocks and also linked to a basal detachment.

It is also important to recognise the propagation directions of faults and to appreciate which faults are

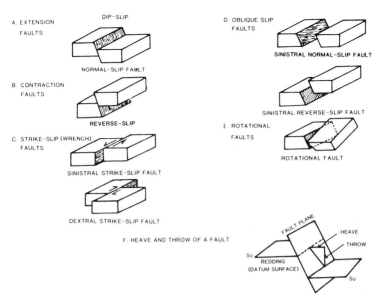

Fig. 6.6 Block diagrams showing the slip (displacement) of faults when the absolute directions of movement can be determined.

Fig. 6.7 Conjugate extension faults developed in interbedded sandstones and siltstones. Field of view 10 m.

89

younger. Above a major detachment fault it has been found that the deformation propagates into the hanging-wall as depicted in Fig. 6.8c. Hanging-wall *roll-over* anticlines (Fig. 6.3a) with antithetic and synthetic faults develop above the listric fault (Fig. 6.8c).

Symbols for recording extensional fault systems on the map are shown in Fig. 6.9, together with an example of the stereographic projection of an extensional fault system. An example of a map of an extensional fault terrane is shown in Fig. 6.10, illustrating the linkage of faults to a basal detachment.

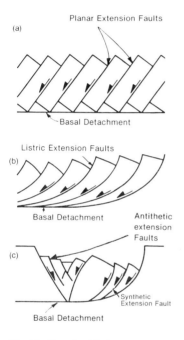

Fig. 6.8 Extensional fault systems: (*a*) domino extensional fault system — linked planar rotational extension faults; (*b*) listric extensional fault system — linked listric extensional faults (with geometrically necessary rotation); (*c*) propagation of second-order faults above a major listric detachment. The second-order faults may be termed *synthetic* if their sense of movement is the same as that of the major (first-order) fault, or *antithetic* if the movement is in the opposite sense.

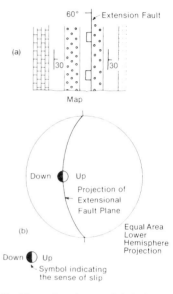

Fig. 6.9 (*a*) General map symbols for dip slip extensional faults. (*b*) Stereographic projection of the relationships shown in a, — note the use of partially filled circles to indicate the downthrown blocks.

6.3.2 *Mapping extension faults*

Structural data that should be collected for extension faults are listed in Table 6.1.

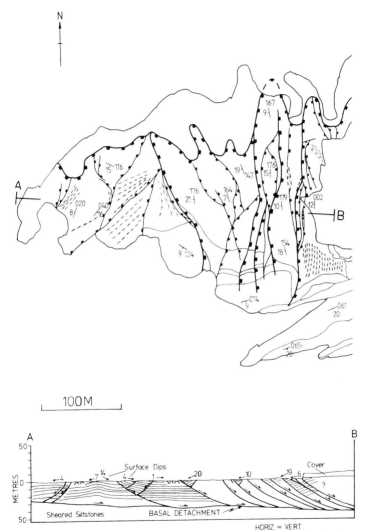

Fig. 6.10 Map of an extensional fault system in deltaic sediments. The cross-section (at the same scale as the map) shows a system of linked listric faults which merge into a basal detachment fault (adapted after a map by Nick Way). Symbols as in Table 2.1 (p. 42).

91

Table 6.1 Data to be collected from extensional faults.

Structure	What to Measure	What Observations to Record	Results of Analysis
	Orientation of fault plane (dip direction) (Figs. 2.5–2.8).	Nature of fault plane: fault rocks. Curvature of fault plane? Width of fault.	Deformation processes. Listric/planar faulting. (Fig. 6.8)
	Orientation of displaced units on both sides of fault (Figs. 2.5–2.8).	Stratigraphic separation. (Fig. 6.2) Sense of movement. Sense of shear.	Displacement direction. (Fig. 6.6) Minimum slip. Amount of extension.
	Lineations on fault plane: grooving, slickensides, slickolites (Figs. 2.10–2.13; Figs. 5.6 & 6.4).	Nature of lineations on fault plane (fibrous slickensides?). (Table 5.3) Movement sense. (Fig. 5.6)	Movement direction. (Fig. 5.6)
		Relationship to other faults. Cross-cutting relationships. Associated folding. (Fig. 3.13).	Fault sequences. (Fig. 6.8) Kinematic development.
	Orientation data on synthetic structures (Fig. 6.8c): faults and fractures (Figs. 2.5–2.8; 2.11–2.13).	Nature of synthetic structures. Movement directions. (Fig. 5.6)	Fault systems. (Fig. 6.6) Movement patterns. (Fig. 6.6) Stress systems. (Fig. 6.1)
	Orientation data on antithetic structures (Fig. 6.8c): faults and fractures (Figs. 2.5–2.8, 2.11–2.13).	Nature of antithetic structures. Movement directions. (Fig.5.6)	Fault systems. (Fig. 6.6) Movement patterns. (Fig. 6.6) Stress systems. (Fig. 6.1)

6.4 Contraction faults

In this section low-angle contraction faults, i.e. *thrust faults* are discussed. Thrust faults are found in most compressional tectonic regimes. Although the complete fault geometry is often not exposed, you should be aware of the geometric consequences of thrusting and recognise its effects on map patterns. Summaries of modern thrust terminology and geometries are given by Boyer and Elliott (1982) and Butler (1982).

6.4.1 Thrust faults in very low grade metamorphic terranes

Many thrust faults in these terranes (e.g. foreland fold and thrust belts) have a staircase geometry, comprising long, bedding-parallel glide zones—*flats*—separated by short, steeper-angled thrusts or *ramps* (Fig. 6.11).

Thrust faults are *three dimensional* and they can be considered to have a *slipped region* surrounded by a *ductile bead* (a cleavage front or anticline–syncline pair). Thus, at thrust fault terminations (*tip line*) the thrust faults die out into an anticline–syncline pair. In three dimensions, a thrust fault may have a complex ramp geometry with *frontal ramps* (perpendicular to the movement direction); *lateral ramps* (parallel to the movement direction); and *oblique ramps* (oblique to the movement direction), (Fig. 6.11d).

Thrust faults may be *linked* by *wrench faults* (tear faults) which root in the underlying *floor or sole thrust* (Fig. 6.11e). Such tear faults occur on all scales linking small imbricate faults, to large thrust systems.

When a thrust sheet moves over a ramp (Fig. 6.11a) it becomes folded, forming the characteristic 'snake's head' structure (Fig. 6.11a) with kink and box-fold structures. Consider a segment that moves over the ramp: it folds, unfolds, folds again and then finally unfolds (Fig. 6.11a). Each stage of the deformation will be accompanied by internal strains and structures, i.e. fractures, with the result that complex superposed structures are produced (i.e. superposed cleavages, fracture patterns).

A thrust sheet that has moved over a ramp will produce an uplifted segment of lower stratigraphy. In three dimensions, this will be bounded by culmination walls (Fig. 6.11e). Erosion of this culmination will produce a tectonic 'window'. Examples of thrust fault geometries are given in Fig. 6.12. It is pertinent to note the development of folds above a thrust fault, and these can give important information on the geometry of the underlying thrust plane.

6.4.2 Thrust faults in higher grade terranes

In low green schist-facies terranes and at higher metamorphic grades, thrust faulting is commonly associated with folding and the development of penetrative foliations. In such situations a staircase fault geometry (Section 6.4.1) may not be well developed and the thrust fault may have a smooth trajectory (Fig. 6.13). Folding may be intimately associated with the thrust faulting with the faults cutting folded beds at high angles and penetrative foliations may be developed (Fig. 6.13).

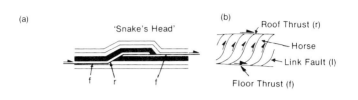

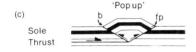

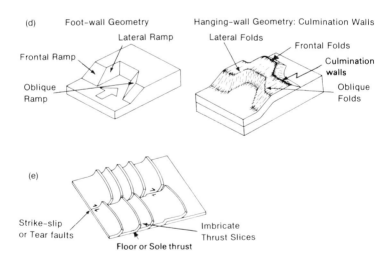

Fig. 6.11 on p. 94 (*a*) Staircase thrust fault trajectory showing the development of *flats* (**f**) where the thrust fault is parallel to the bedding and *ramps* (**r**) where the thrust fault cuts up-section through the bedding. Note the development of a typical '*snake's head*' type structure as the hanging-wall thrust sheet moves up and over the ramp. Note also the development of geometrically necessary folds over the ramp structures. (*b*) Development of a *duplex* where stratigraphy is duplicated by repetition on *link thrusts* (**l**) which link to a *floor* thrust (**f**) and to a *roof* thrust (**r**). Individual thrust segments are termed *horses*. (*c*) Development of a *back thrust* (**b**) and a forward propagating thrust (**fp**) resulting in an uplifted triangle zone. (*d*) The three-dimensional nature of thrust fault surfaces. In the foot-wall sheet there will be *frontal, oblique* and *lateral* ramps. In the hanging-wall sheet there will be geometrically necessary *folds* associated with the ramps – e.g. *frontal folds, oblique folds* and *lateral folds*. (*e*) Thrust sheets linked by strike-slip faults (tear faults).

Fig. 6.12a Thrust fault in the Moine Thrust zone of NW Scotland. Older dark siltstones are thrust over younger light-coloured dolomites along a planar thrust fault.

Fig. 6.12b Curved duplex link fault in the Moine Thrust zone of NW Scotland.

Fig. 6.12c Small-scale duplex in limestones in the Canadian Rocky Mountains. Note the smooth trajectories of the floor (**f**) and roof thrusts (**r**) and the sigmoidal link thrusts (**l**) between them.

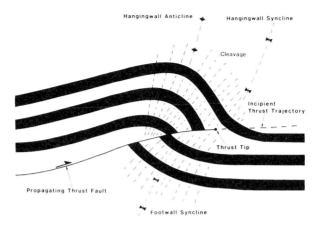

Fig. 6.13 Smooth thrust fault trajectories. Folding occurs at the same time as the thrust faulting, giving rise to high cut-off angles and associated cleavages at the thrust tip.

An example of a map of a thrust terrane is shown in Fig. 6.14. Here there is an intimate association of folds and thrusts which cut the strata at high angles. This is also reflected in the smaller scale structures (Fig. 6.15) where the thrusts cut bedding at a high angle (90°), thus indicating that the thrusting accompanied or post-dated the folding.

6.4.3 Basic rules for thrust faults

Some basic rules governing the geometry and the kinematics of thrust faults can now be formulated to assist in the mapping of these structures:

1 Thrusts bring older rocks over younger rocks, unless they develop in already folded strata.
2 Thrusts climb up stratigraphic

section, unless they develop in already folded strata.
3 Thrusts propagate in the direction of movement.
4 In a thrust system, topographically higher but older thrusts are carried 'piggy back' on lower, younger thrusts.
5 Higher (older) thrusts are folded as lower, younger thrusts climb ramp structures.
6 Ramp angles or 'cut-off' angles are generally between 15° and 30° to the bedding datum.

These basic rules have been well proven in many thrust belts, but may be invalidated if it can be shown that a later thrust may have cut through earlier formed structures from the rear. Such a thrust fault is termed an *out of sequence thrust* and rules 1–6 may be invalidated by these thrusts.

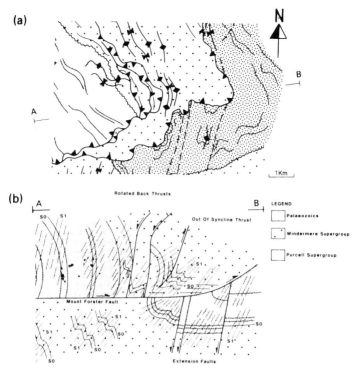

Fig. 6.14 (a) & (b) An example of a thrust terrane in southern British Columbia, illustrating (a) a close association of folding and thrusting and (b), a sketch section.

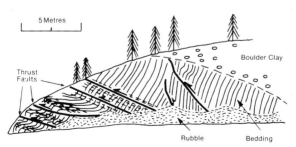

Fig. 6.15 Detailed field sketch of thrusting in the Purcell Mountains, Canada, showing high thrust cut-off angles; folding and thrusting occurred together.

6.4.4 Movement direction of thrust faults

Movement on thrust faults can be determined by:

1 *'Bow and arrow rule'*: In plan, thrust faults are commonly curved (Fig. 6.14a) and the movement direction is generally normal to the 'string' formed by connecting the ends of the 'bow', i.e. in the direction of the 'arrow'.
2 Movement is normal to frontal ramps in the thrust fault.
3 Movement is normal to folds produced over frontal ramps or folds produced in the ductile bead ahead of the thrust tip line.
4 Movement is parallel to lateral ramp systems and associated folds.
5 Movement can be determined from the development of slickensides and other lineations on brittle fault planes.
6 Movement can be determined by ductile lineations in or near the fault plane—*cf.* mylonitic lineations (Fig. 5.3).
7 In ductile thrust regimes, folds will form initially parallel to the thrust front, but then subsequent deformation will rotate them into parallelism with the transport direction.

6.4.5 Mapping thrust faults

Data that should be collected from thrust faulted terranes are shown in Table 6.2.

6.5 Strike-slip or wrench faults

Strike-slip or wrench faults are vertical (in the upper part of the crust) and have horizontal movement directions (Figs. 6.1, 6.2 and 6.6). They are classified as *dextral* (right lateral) or *sinistral* (left lateral) (Fig. 6.2 and 6.6).

The following features of wrench fault systems have been found:

1 Wrench faulting is initiated (Fig. 6.16a) by the development of Riedel shears (R_1 and R_2) oriented at $\simeq 30°$ to the maximum principal compressive stress σ_1. Movement is concentrated on one Riedel system R_1 which is synthetic to the direction of maximum resolved shear stress (e.g. Fig. 6.16b) and the antithetic system R_2 is subordinate. In some systems synthetic P, and antithetic X shear faults may also develop.
2 The major through-going wrench fault is oriented at 45° to the maximum compressive stress σ_1.
3 Secondary wrench faults (antithetic and synthetic) occur along the Riedel shear directions and these may in turn develop their own secondary fault patterns.

Fig. 6.16a represents the deformation on a wrench fault system in terms of a 2D deformation ellipse which shows not only the antithetic and synthetic wrench faults but also zones of compression and extension within the system. Zones of compression can give rise to folds and

Table 6.2 Data to be collected from contractional faults.

Structure	What to Measure	What Observations to Record	Results of Analysis
	Orientation of fault plane (dip direction) (Figs. 2.5–2.8).	Nature of fault plane: fault rocks. (Fig. 6.18, Table 6.4) Curvature/stepped nature of fault plane? Width of fault zone. (Fig. 6.11)	Deformation processes. Listric/planar/stepped fault. (Fig. 6.11a)
	Orientation of displaced units on both sides of fault (Figs. 2.5–2.8).	Stratigraphic separation/overlap. Sense of movement. Sense of shear.	Displacement direction. (Fig. 5.6) Minimum slip. Amount of contraction.
	Lineations on fault plane: grooving, slickensides, slickolites (Figs. 2.4–2.13, 5.6 and 6.4).	Nature of lineations on fault plane (fibrous slickensides?) (Table 5.3) Movement sense. (Fig. 5.6)	Movement direction. (Fig. 5.6)
		Relationships to other faults. Cross-cutting relationships: Imbricate fan? duplex? out of sequence? Ramps? Associated folding. (Figs. 6.11a and 3.13)	Fault sequences. Kinematic development.
	Orientation data on synthetic structures: faults and fractures (Figs. 2.5–2.8, 2.11–2.13),	Nature of synthetic structures. Movement directions. (Fig. 5.6)	Fault systems. Movement patterns. Stress systems. (Fig. 6.1)
	Orientation data on antithetic structures: faults and fractures (Figs. 2.5–2.8, 2.11–2.13).	Nature of antithetic structures. Movement directions. (Fig. 5.6)	Fault systems. Movement patterns. Stress systems. (Fig. 6.1)

thrust faults, whereas zones of extension develop extension (or normal) faults.

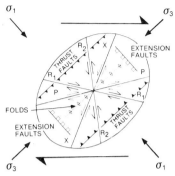

Fig. 6.16a Regional strain ellipse associated with a wrench/strike-slip fault system. The Reidel shear faults are the synthetic R_1 and antithetic R_2 systems (in most cases displacements are minor on these faults). In some systems synthetic P and antithetic X shears may also develop. Folds and contraction (thrust) faults are developed at 90° to σ_1, whereas extension (normal) faults are developed 90° to σ_3.

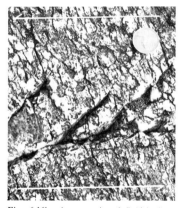

Fig. 6.16b An example of Reidel shears associated with a sinistral strike-slip fault in quartzites.

6.5.1 En-echelon wrench faults

Wrench faults commonly occur in en-echelon systems and, as such, can be termed right-stepping or left-stepping (Fig. 6.17). Types 1 (right-stepping) generates an extensional zone (normal faults) whereas Type 2 (left-stepping) generates a compressional zone (folds and reverse faults). *En-echelon folding* typically accompanies wrench faulting. These styles have been well demonstrated in high-level brittle structures and the resulting geometric patterns should be taken into account during mapping programmes (Fig. 6.17).

6.5.2 Movement directions

Movement directions on wrench fault systems will generally be horizontal–sub-horizontal. They can be determined by:

1 Slickensiding or grooving on exposed planes; (Fig. 6.4)
2 Analysis of patterns of antithetic and synthetic faulting, and of minor fracture patterns; (Fig. 6.16a)
3 Rotation of structures into the fault zone, indicating movement directions; (Fig. 6.16b)
4 Development of horizontal linear fabrics in rocks adjacent to the fault zone.

6.5.3 Mapping wrench faults

Table 6.3 lists the data that should be collected from wrench fault zones.

Table 6.3 Data to be collected from wrench faults.

Structure	What to Measure	What Observations to Record	Results of Analysis
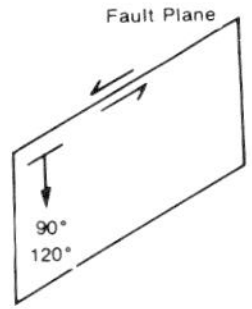	Orientation of fault plane (dip direction) (Figs. 2.5–2.8 and 6.1).	Nature of fault plane fault rocks. (Fig. 6.18 and Table 6.4) Width of fault zone.	Deformation processes.
	Orientation of displaced units on both sides of fault (Figs. 2.5–2.8).	Movement direction. (Fig. 5.6) Sense of shear.	Displacement direction. (Fig. 5.6) Amount of slip/offset.
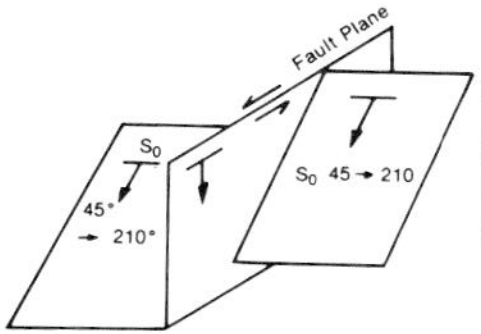	Lineations on fault plane: grooving, slickensides, slickolites (Figs. 5.6, 6.4; 2.11–2.13).	Nature of lineations on fault plane (fibrous slickensides?). (Table 5.3) Movement sense. (Fig. 5.6)	Movement direction. (Fig. 5.6)
		Relationships to other faults. Cross-cutting relationships. Associated folding.	Fault sequences. (Figs. 6.16 and 6.19) Kinematic development. (Fig. 6.16a)
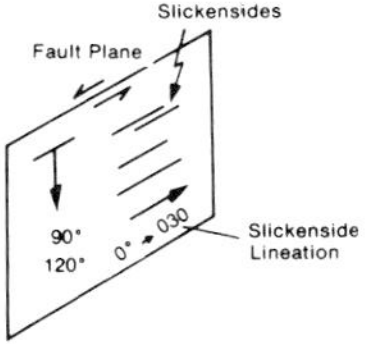	Orientation data on synthetic structures Riedel shears R_1, R_2, P shear (Fig. 6.16a). Second and third order faults Associated folds (Fig. 6.16a).	Nature of synthetic structures. (Fig. 6.16a) Movement directions. (Fig. 5.6)	Fault systems. (Fig. 6.16a) Movement patterns. (Fig. 6.16a) Stress systems. (Fig. 6.16a)

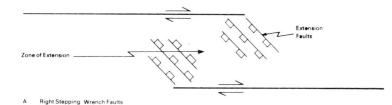

A Right Stepping Wrench Faults

B Left Stepping Wrench Faults

Fig. 6.17 En-echelon wrench faults showing different senses of off-set. (*a*) Right-stepping faults (going from the upper fault to the lower fault) generate a zone of extensional faulting. (*b*) Left-stepping faults generate a zone of compression: thrust faulting and folding.

6.6 Fault rocks

When mapping faults it is appropriate to carefully describe the types of fault rock developed as these may provide important clues as to the conditions of deformation during faulting, e.g. *gouge and breccia* commonly form at high crustal levels, whereas *mylonites* commonly form at deeper crustal levels (at higher temperatures and pressures) and indicate plastic deformation. Sibson (1977) has provided a classification for describing fault rocks (Table 6.4).

The classification shows three main subdivisions:

1 *Incohesive unfoliated fault rocks*
   Fault gouge — powdered, crumbly fault rock (falls apart in your fingers) — with less than 30% visible rock fragments (Fig. 6.18a).
   Fault breccia — fault rock consisting of loose or loosely bound angular rock fragments often in a gouge matrix (Fig. 6.18a).

2 *Incohesive foliated fault rocks* —
   Foliated fault gouge and foliated fault breccia — gouge and breccia as described above but having a distinct planar fabric in the fault gouge and alignment and preferred shape orientation of the breccia fragments.

3 *Cohesive unfoliated fault rocks* —
   Pseudotachylite — a frictional melt generated by fault movement. Characteristically black and

Table 6.4 A classification of fault rocks (modified after Sibson, 1977).

		Random–fabric		Foliated	
Incohesive		*Fault breccia* (visible fragments > 30% of rock mass)		*Foliated fault breccia*	
		Fault gouge (visible fragments < 30% of rock mass)		*Foliated gouge*	
Cohesive	glass/devitrified glass		Pseudotachylite	*Foliated Pseudotachylite*	
	Nature of matrix — tectonic reduction in grain size dominates; grain growth by recrystallisation and neomineralisation		Crush breccia Fine crush breccia Crush microbreccia	(fragments > 0·5 cm) (0·1 cm < fragments < 0·5 cm) (fragments < 0·1 cm)	0–10%
		cataclasite series	Protocataclasite	Protomylonite	10–50%
			Cataclasite	Phyllonite varieties / Mylonite (mylonite series)	50–90%
			Ultracataclasite Flinty crush rock	Ultramylonite	90–100%
	grain growth pronounced		?	Blastomylonite	

(Proportion of matrix)

glassy and occurring in veins and seams with other fault rocks (Fig. 6.18b).

Crush breccia—a hard, intact, unfoliated rock consisting of angular fragments with no preferred orientation (Fig. 6.18c).

Cataclasites—intact and unfoliated rocks with the grain size tectonically reduced by fracturing. Cataclasites vary from *Protocataclasites*—highly fragmented rock showing many large fragments of the original rock type to *Ultracataclasites*—dark ultra fine-grained almost glassy rocks with no relicts of the original rock type.

4 *Cohesive foliated fault rocks*—

Mylonite series—generally fine-grained dark foliated rocks with ductile fabrics (e.g. folds) and grains showing a reduction in size by plastic processes (Fig. 6.18d).

Phyllonite series—mica-rich mylonites which have the silky appearance of phyllites and a well developed foliation (Fig. 6.18e).

Fault rocks should be described in the field using Table 6.4. Samples should be collected for petrological examination to confirm the field interpretations.

Shear zones

Shear zones are narrow, sub-parallel-sided zones of strong non-coaxial deformation. They occur on all scales from crustal size to microscopic and range from brittle to ductile in character—in fact, many fault zones can be treated as shear zones. Brittle shear zones form in the upper 5 km of the crust, whereas ductile shear zones generally form below 5–10 km in the crust. Ductile shear zones are common in deformed crystalline basement rock. They are characterised by high shear strains, strong foliation development and large displacements (relative to their width). Typically they form in homogeneous isotropic rocks, but once formed, deformation is concentrated within the shear zone.

6.7 Geometry of shear zones

Shear zones can form conjugate arrays, and these, or the individual shear zones, can be analysed to determine strain displacements and palaeostress directions. Ramsay (1982) tabulated the properties of shear zones in the crust (Table 6.5).

6.7.1 Types of shear zones

The geometries of simple brittle-to-ductile shear zones are shown in Figs. 6.19–6.21. In each case simple shear (Ramsay, 1967) is assumed, and the *shear zone boundaries* are at 45° to the principal compressive stress σ_1.

Brittle shear zone Three sets of fractures may develop in the shear (fault) zone. R_1 principal Reidel shears; R_2 conjugate Riedel shears (generally subordinate); and P synthetic shears, whose directions are imposed by boundary conditions and may or may

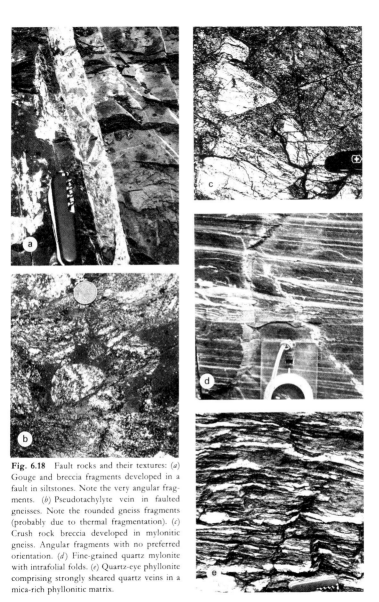

Fig. 6.18 Fault rocks and their textures: (*a*)
Gouge and breccia fragments developed in a
fault in siltstones. Note the very angular fragments. (*b*) Pseudotachylyte vein in faulted
gneisses. Note the rounded gneiss fragments
(probably due to thermal fragmentation). (*c*)
Crush rock breccia developed in mylonitic
gneiss. Angular fragments with no preferred
orientation. (*d*) Fine-grained quartz mylonite
with intrafolial folds. (*e*) Quartz-eye phyllonite
comprising strongly sheared quartz veins in a
mica-rich phyllonitic matrix.

not develop. Stress orientations and sense of shear orientations can be deduced from the pattern of Riedel shears and from the fabric in the fault gouge (Fig. 6.19a). An example of a brittle shear zone is shown in Fig. 6.19b.

Semi-brittle shear zone (en-echelon tension gashes) Here the deformation is partly ductile with the development

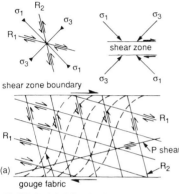

Fig. 6.19 Brittle shear zones: (*a*) Brittle shear zone showing the development of a gouge fabric, R_1 and R_2 Riedel shears and a low-angle P shear. The stress systems for the Riedel shear systems and for the through-going shear zone are also shown. (*b*) Brittle dextral shear zone in massive volcanic breccias. Note the development of R_1 Riedel shears.

of pressure-solution cleavage and partly brittle with extensional veins developed (total volume change = 0). The tension gashes have their tips oriented parallel with σ_1 and are generally infilled with fibrous minerals that grow incrementally in the σ_3 direction (Fig. 6.20a). The pressure-solution cleavage (if developed) forms 90° to σ_1 and the vein tips, but becomes rotated towards parallelism with the shear zone walls in the central part of the shear zone. An example of a semi-brittle shear zone with en-echelon tension gash veins is shown in Fig. 6.20b.

Ductile shear zone Here the deformation is entirely ductile and produces a strong schistosity which originates at 45° to the shear zone (and perpendicular to σ_1). As deformation proceeds the schistosity is rotated towards the shear zone plane until, at large strains, it is nearly parallel to the shear zone boundaries (Fig. 6.21a). An example of a ductile shear zone is shown in Fig. 6.21b.

The total shear strain and displacement within ductile and brittle-ductile simple shear zones are easily analysed using the methods of Ramsay and Graham (1970) but require detailed grid mapping and/or photography so that all the structural elements can be recorded *across* the shear zone.

Conjugate shear zones Shear zones may develop in conjugate arrays (Fig. 6.22) and, as such, may be analysed to determine principal stress orientations (Table 6.5).

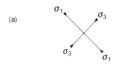

Note: It is important to remember
that a shear zone must start and end.
At the ends of a shear zone, complex
foliation and strain patterns occur, so
that the simple geometries described

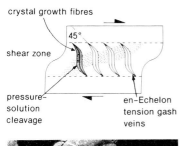

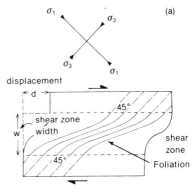

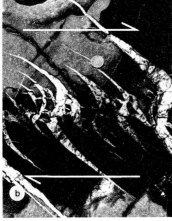

Fig. 6.20 Semi-brittle to semi-ductile shear zones: (*a*) Semi-brittle shear zone showing the development of en-echelon fibrous tension gashes. The fibre orientations reflect the incremental orientations of the σ_3 stress axis as the tension gash grows. Inside the shear zone a pressure-solution cleavage may be locally developed. The ideal orientation of the σ_1 and of the σ_3 stresses outside the shear zone are also shown. (*b*) Semi-brittle dextral shear zone in greywackes. Several sets of en-echelon quartz tension gashes are developed.

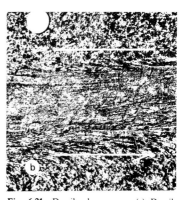

Fig. 6.21 Ductile shear zones: (*a*) Ductile shear zone showing the development of a foliation at 45° to the shear zone margin (and 90° to σ_1) and rotation of this foliation into the shear zone. (*b*) Ductile sinistral shear zone in a tonalite. Note the development of a schistosity at the shear zone margins, and the rotation of this schistosity into parallelism with the shear zone.

Table 6.5 Geometric properties of shear zones in the crust.

Approximate depth	Metamorphic facies	Structural features of shear zones	2θ, angle between conjugate shear zones
>10 km ductile shear zones	granulite, amphibolite blueschist	ductile flow, strong sigmoidal schistosity in zones.	120°–90°
5–10 km ductile, brittle-ductile shear zones.	greenschist, zeolite	ductile to semi-brittle; localised schistosity; en-echelon vein arrays; pressure-solution features.	90°–60°
0–5 km brittle shear zones	Anchimetamorphism, no metamorphism	Brittle; fault breccia and clay gouge; some pressure-solution features.	60°

above only apply in the *central* part of a shear zone undergoing *simple shear* deformation.

Fig. 6.22 A conjugate array of semi-brittle shear zones as seen on a bedding plane in deformed sandstones. Note the en-echelon quartz tension gashes and pressure-solution seams at 90° to the tension gashes. Field of view 2 m.

108

6.8 Structures in shear zones

The orientation of structural elements outside and within the shear zone gives important information on the *sense of shear, strain and displacement* within the shear zone, and a brief summary of the relevant structures to be observed in the field is given below.

Planar structures: Foliation Initiates at 45° to the shear zone and with increasing strain is rotated towards parallelism with the shear zone walls (Fig. 6.21b).

Passive layers: These are layers of rock or pre-existing foliations which have no contrast in competency with, or mechanical effect upon the shear zone and are simply rotated into the shear zone (Fig. 6.23a).

Active layers: These have a competency contrast with the shear zone material, and are folded or boudinaged according to their initial orientation (Fig. 6.23b). Fold axes will generally not be in the XY plane of the strain ellipsoid for the shear zone.

Linear structures: Lineations Many shear zones develop a strong mineral stretching lineation parallel to the shear direction (e.g. Fig. 6.21a). Pre-existing lineations (e.g. pre-existing fold axes) are rotated towards parallelism with the shear direction. In such situations the fold hinge lines

Fig. 6.23a Dextral shear zone in psammitic schists showing the rotation of a pre-existing foliation into parallelism with the shear zone boundaries (horizontal).

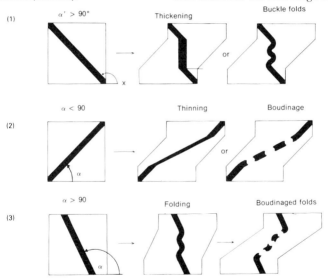

Fig. 6.23b The effects of shear zone deformation on competent layers of different initial orientations: (1) a layer at an initial angle α, which lies at more than 90° to the shear zone boundary, becomes thickened and folded; (2) layer at an angle α which initially is <90° to the shear zone boundary, and becomes thinned and boudinaged; (3) layer at an angle α initially >90° to the shear zone boundary (less than in a above) and which first shortens by folding and is subsequently stretched and boudinaged, (reproduced with permission from Ramsay, 1980).

are strongly curved and sheath folds are produced (Fig. 6.24), typically with 'eye'-shaped cross-sections.

6.8.1 Sense of shearing in shear zones

The correct evaluation of the sense of shear in shear zones is vital in structural mapping, e.g. to determine whether a mylonite/shear zone is extensional or contractional.

The field criteria that may be used (with care) as kinematic indicators to deduce sense of shear are listed below.

1 En-echelon tension gashes (Fig. 6.20);
2 Orientations of foliation (Fig. 6.21);
3 Orientations of gouge and Riedel shears (Fig. 6.19);
4 Asymmetric augen structures (Fig. 6.25a);
5 Broken and displaced pebbles, grains and crystals (Fig. 6.25b);
6 C and S fabric relationships (Figs. 6.26a & b); (C = shear surface sub-parallel to shear zones; S = schistosity surface). The orientation of S surfaces which occur between the shear surfaces is antithetic to the sense of shear, i.e. against it.
7 Development of shear bands in homogeneous, strongly foliated rocks (Figs. 6.27a and c).

Fig. 6.24 Curved fold hinge lines and folds with eye-shaped cross-sections — 'sheath folds'. These are formed and deformed in a shear zone.

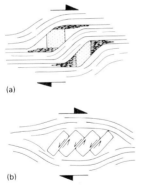

(a)

(b)

Fig. 6.25 Kinematic indicators for simple shear deformation. (a) Sigmoidal feldspar porphyroblasts with tails of recrystallised feldspar trailing into the foliation and producing asymmetric augen structures. (b) Fragmented augen showing antithetic slip on fracture planes in the grain, (see Simpson and Schmid, 1983, for details).

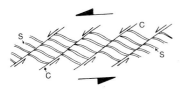

Fig. 6.26a The geometry of C and S fabrics as seen in shear zones. The C plane is the shear plane and S is the schistosity plane. The C plane has an extensional geometry with respect to the sense of shear. (Fig. 6.26b on p. 112)

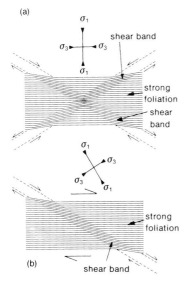

Fig. 6.27a Shear band development in strongly foliated rocks, e.g. mylonites. (*a*) flattening produces conjugate low-angle shear bands which are approximately 30° to the mylonitic foliation; (*b*) simple shear geometry produces a single shear band orientation — sometimes called an extensional crenulation cleavage. Shear bands are important indicators of the sense of shear in mylonitic rocks. (Fig. 6.27b on p. 112)

Note: The sense of shear is best determined using criteria 1,2,3,6 and 7. Criteria 4 and 5 need to be very carefully examined and many observations made before the sense of shear can be reliably ascertained. In addition to the mesoscopic fabric elements which allow determination of the sense of shear, microscopic analysis may also allow the sense of shear to be analysed. Shear zones or mylonitic foliations should therefore be sampled (oriented samples, section 2.7) for laboratory analysis.

6.9 Mapping shear zones

Where possible the factors listed in Table 6.6 should be measured and recorded.

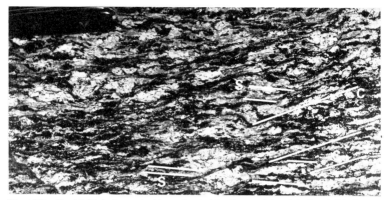

Fig. 6.26b C and S fabrics in a mylonitic granite. A sinistral sense of shear is indicated. C is the shear plane and S is the schistosity plane. (Fig. 6.26a on p. 111)

Fig. 6.27b Low-angle shear bands developed in strongly foliated melange (foliation horizontal). (Fig. 6.27a on p. 111)

Table 6.6 Data to be collected from observations on shear zones.

Structure	What to Measure	What Observations to Record	Results of Analysis
Brittle shear zone 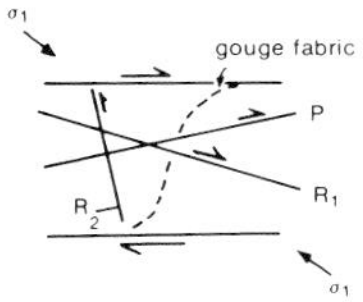	Orientation of shear zone boundaries (Figs. 2.5–2.8). Orientation of Riedel fractures R_1 and R_2 (Fig. 6.19, & Figs. 2.5–2.8). Orientation of fabric in fault gouge. (Fig. 6.8) Orientation of P fracture (if developed). (Fig. 6.19) (Additional data — orientation of structures outside shear zone). (Fig. 6.23)	Nature of shear zone. (Fig. 6.19) Width of shear zone. Fault rocks developed. (Fig. 6.18, Table 6.4) Veining and/or pressure-solution. (Fig. 6.23 and 7.6b) Fracture orientations relative to shear zone. (Fig. 6.19a) Sense of shear, (Fig. 6.19a) Displacement. Structures outside shear zone.	Stress systems. (Fig. 6.19a) Sense of shear. (Fig. 6.19a) Displacement. Deformation processes.
	Orientation data on conjugate array. (Fig. 6.22)	Observations on conjugate array. (Table 6.5)	Stress systems. (Fig. 6.19a) Sense of shear. (Fig. 6.19a) Displacement. Deformation processes.

113

Table 6.6 Cont'd

Structure	What to Measure	What Observations to Record	Results of Analysis
Semi-brittle shear zone (en-echelon tension gashes) 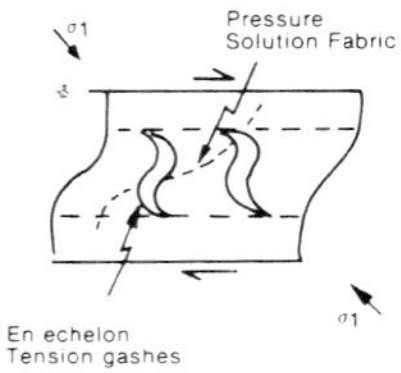	Orientation of shear zone boundaries (Figs. 6.20 & 2.5–2.8). Orientation of crack tips. (Fig. 6.20a) Orientation of intersection of crack tips with shear zone boundary. (Fig. 6.20a) Orientation of pressure-solution fabric at shear zone margins. (Fig. 6.20a)	Nature of shear zone. (Figs. 6.20, 6.22) Width of shear zone. Nature of veins— fibrous or massive. (Figs. 6.20a, 7.6) Nature of foliation in shear zone. (Figs. 4.1, 4.2) Sense of shear. (Fig. 6.20a) Displacement. Photograph/sketch of shear zone. Structures outside of shear zone.	Stress systems. (Fig. 6.20a) Sense of shear. (Fig. 6.20a) Displacement. Strain in shear zone. Deformation processes.
	(Additional data on orientation of structures outside shear zone). Orientation data on conjugate array. (Fig. 6.22)	Observations on conjugate array.	Stress systems. (Fig. 6.20a) Sense of shear. (Fig. 6.20a) Displacement. Strain in shear zone. Deformation processes.

114

Ductile shear zone

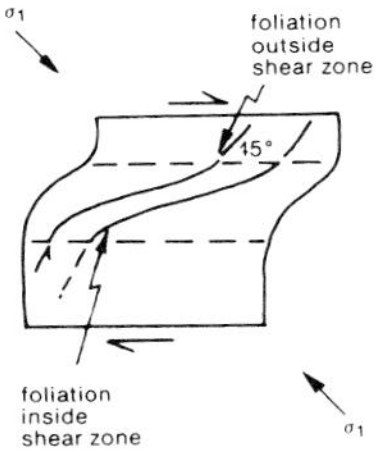

Orientation of shear zone boundaries (Figs. 6.21 & 2.5–2.8).	Nature of shear zone. (Fig. 6.21a)	Stress systems. (Fig. 6.21a)
Orientation of foliations at shear zone boundaries. (Fig. 6.21)	Width of shear zone.	Strain distribution.
Orientation of lineations in shear zone (ML). (Figs. 2.11 to 2.14)	Nature of foliation. (Figs. 4.1, 4.2)	Sense of shear. (Fig. 6.21a, 6.26, 6.27)
Orientation/vergence of folds in shear zone. (Fig. 3.9)	Sense of shear.(Figs. 6.21, 6.26, 6.27)	Displacement.
Strain of deformed objects across shear zone. (Appendix III)	Displacement.	Deformation processes.
	Nature of folds/vergence. (Fig. 3.9)	
(Additional data on orientation of structures outside shear zone).	Strain in deformed objects. (Appendix III)	
	Photograph/sketch of shear zone.	
	Structure outside shear zone.	
Orientation data on conjugate array. (Fig. 6.22)	Observations on conjugate array. (Table 6.5)	Stress systems. (Fig. 6.21a) Strain distribution. Sense of shear. (Fig. 6.21a, 6.26, 6.27) Displacement. Deformation processes.

115

7

Joints, veins and stylolites

Joints are regular arrangements of fractures along which there has been little or no movement. They are the most commonly developed brittle structures. Veins are fractures infilled with remobilised minerals (e.g. quartz or carbonate). Stylolites — surfaces of dissolution — are included in this Chapter because the development of tectonic stylolites is commonly associated with joint and vein development.

Measurements of the orientations of joints, veins and stylolites are made using the techniques for planar structures, as outlined in Chapter 2. Careful observations are needed in order to determine the type of joint, vein or stylolite, and in particular, to determine the relative age relationships between various joints, veins and stylolites.

7.1 Types of joints

Three basic types of joints are found (Fig. 7.1):

1 *Dilational joints* extensional joints with the fracture plane normal to the least principal stress σ_3 during joint formation (Figs. 7.1a & b).

2 *Shear joints* — often conjugate, enclosing angles of 60° or more. The joint planes may show small amounts of shear displacement (Figs. 7.1c & d).

3 *Combined shear and extension joints* — termed *hybrid* joints which show components of both extension and shear displacement (Figs. 7.1e & f).

4 *Irregular extension joints* are those in which extension occurs in all directions (often due to hydraulic fracturing as a result of high pore fluid pressures). This gives rise to irregular joint patterns (Fig. 7.1g).

Fig. 7.1 Joint types: (*a*) Dilational (extensional) joints. (*b*) Horizontal extensional joint system in sandstones joined by vertical cross-joints producing H and T intersection patterns. (*c*) Shear joints. (*d*) Conjugate shear joints (**s**) intersecting at 75° with coeval, planar, vertical extension joints (**e**) and joined by less regular cross-joints (**c**). (*e*) Hybrid joints — both dilational (extensional) and shear movement. (*f*) Massive volcanics with a set of hybrid joints that intersect at 44°, forming an X pattern. (*g*) Polygonal joint pattern in siltstones, indicating extension in all directions under hydrostatic stress conditions — high pore fluid pressures.

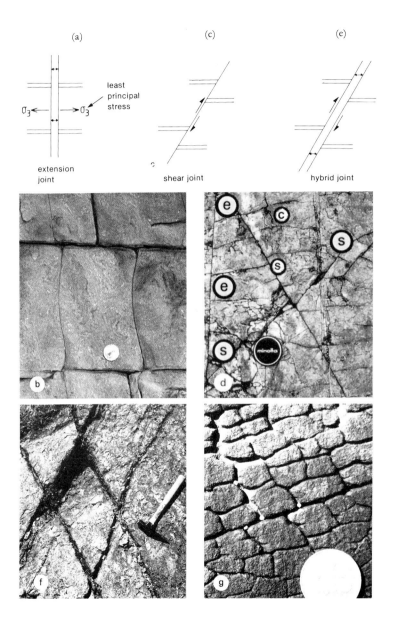

(a) ... least principal stress ... σ_3 ... σ_3 ... extension joint

(c) ... shear joint

(e) ... hybrid joint

If joints are planar, and parallel to sub-parallel so that they form sets, the joints are said to be *systematic*. Joints which can be traced tens and even hundreds of metres are termed *master joints*. Joints which are an order of magnitude smaller but still well-defined, are called *major joints*. Smaller joints occur at all scales down to microscopic.

The frequency of jointing is closely related to bed thickness and lithology — thinner beds have closely spaced joints, more competent beds tend to develop more widely spaced joints.

7.1.1 Analysis of joints

1 Extension joints can be simply analysed by plotting the trace of the joint plane and its pole on a stereographic projection. The direction of σ_3 is the pole to the joint plane which contains the σ_1 and σ_2 axes (Fig. 7.2a). Extension joints alone will not give σ_1 and σ_2 orientations — other dynamic indicators such as vein fibre orientations are needed.

2 Shear joints commonly form conjugate arrays whose angle of intersection (α) is greater than 60°. Plotting these on the stereographic projection (Fig. 7.2b) shows that the line of intersection of the planes is the σ_2 axis. σ_1 bisects the acute angle between the joint planes and σ_3 is at 90° to both σ_1 and σ_2 (Fig. 7.2b).

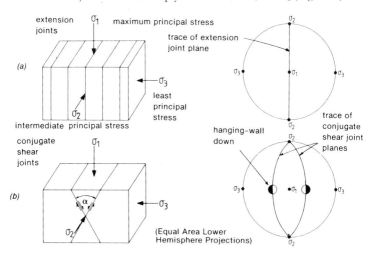

Fig. 7.2 Dynamic (stress) analysis of joint systems: (*a*) Extension joints showing development 90° to σ_3; (*b*) Shear joints showing that σ_1 bisects the acute angle formed by the intersection of conjugate joint planes. The stereographic projections show the projections of the joint planes, the projections of the principal stress directions and the movement directions on the shear joints — solid half circle indicates hanging-wall down.

Fig. 7.3a Plume structures on a joint surface in glacial clays.

Fig. 7.3b Rib marks on a jointed sandstone surface.

3 Age relationships between joints: younger joints generally abutt against and do not cut older joints. Typically T or H patterns result (Fig. 7.1b) with the younger joint (the upright of the T or the cross-bar of the H) abutting against the older joints.

7.1.2 Joint surfaces

Surface markings on joints are commonly either herringbone or plumose marks (Fig. 7.3a) and are thought to indicate extensional joints (Hancock, 1985) but the significance of rib marks (Fig. 7.3b) remains uncertain.

7.2 Joints in fold and fault systems

Joint systems are often arranged symmetrically in fold and fault systems. Here we briefly summarise typical joint patterns in folded rocks.

Cylindrical folds (Section 3.1) typically may exhibit the joint and meso-scale fault systems shown in Fig. 7.4. Extensional fracture systems are commonly parallel or normal to the fold axis (Fig. 7.4a & b). Contractional structures such as small thrust faults and stylolites in a fold system are shown in Fig. 7.4c. Shear joint systems are commonly developed on the fold limbs.

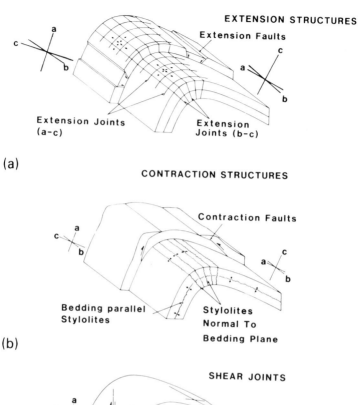

(a)

EXTENSION STRUCTURES

Extension Faults

Extension Joints
(a–c)

Extension
Joints (b–c)

CONTRACTION STRUCTURES

Contraction Faults

Bedding parallel
Stylolites

Stylolites
Normal To
Bedding Plane

(b)

SHEAR JOINTS

Shear
Joints

Shear
Joints

(c)

Fig. 7.4 Joint, fracture, minor fault patterns and stylolites developed in a cylindrical fold system (note that the tectonic axes *a*, *b* and *c* vary around the fold): (*a*) Extensional structures. Some joints are infilled with vein quartz. (*b*) Contractional structures. (*c*) Shear joints and fractures.

Non-cylindrical fold systems (Fig. 3.7) have a fracture architecture that is controlled by the variation of the stress system and hence slip direction *around the fold*. This is illustrated in Fig. 7.5a and an example shown in Fig. 7.5b.

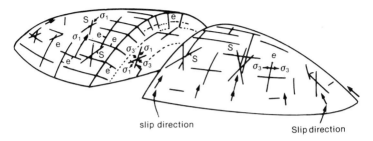

slip direction

Slip direction

S Shear joints
e Extension joints

Fig. 7.5a Fracture patterns developed around a non-cylindrical fold. Note that the conjugate shear fractures (s) and extension fractures (e) vary with slip direction around the fold.

Fig. 7.5b Example of a non-cylindrical fold in siltstones, showing development of shear fractures (s) and extension fractures (e).

7.3 Veins

Many complex forms of veins are found, but those which form regular arrays can be measured and analysed in a similar fashion to joints. Many veins contain growth fibres which record incremental movements as the vein formed. This aids the identification of vein systems which can be classified into extensional, shear and hybrid systems (similar to joints Fig. 7.1).

1 Extensional veins form normal to σ_3 and have fibres perpendicular to the vein walls.
2 Shear or hybrid vein systems have fibres that are oblique to the vein walls (Fig. 7.6).

3 The fibre axis in fibrous veins is approximately parallel to the σ_3 orientation at any stage during the fibre growth. Hence curved fibres in undeformed veins reflect the change in vein orientation with respect to the σ_3 axis (Fig. 6.20).
4 En-echelon vein systems (Fig. 6.20b) are commonly found in semi-brittle shear zones where they can be used to analyse the kinematics and displacements of the shear zones.

Veins indicate high, albeit transient, pore fluid pressures during deformation and are commonly associated with pressure-solution seams (Fig. 7.6b). Vein orientations,

Fig. 7.6a Hybrid vein (shear and extensional components of movement) showing the development of fibrous quartz crystals as an infilling.

Fig. 7.6b Quartz veins in a sinistral shear zone showing characteristic sigmoidal shape and cut by dark pressure solution seams.

122

architecture and age relationships should be measured and analysed in a similar fashion to that described for joints, with the additional measurement and analysis of the *fibre orientation* in the vein system. Veins should be sampled if detailed structural studies are being carried out. Analysis of fibre growth in veins will allow the displacement history to be determined (oriented samples are needed) and fluid inclusions in undeformed vein minerals will give important information on the temperature of formation of the veins.

7.4 Stylolites

Stylolites (Fig. 7.7) are surfaces of dissolution associated with contractional or shear strains (e.g. Fig. 7.4c). They indicate volume loss and may form parallel/sub-parallel to bedding during burial (compactional strains). Tectonic stylolites may form at high angles to bedding during folding, indicating a component of layer-parallel shortening.

Stylolites are commonly associated with joints and veins, and should be measured and analysed with them. They are found in many rock types, including sandstones, and commonly occur in fine-grained carbonates. The stylolitic seam often appears dark and contains a residue of insoluble material (carbonaceous matter, clay and ore minerals) and in places, low-temperature metamorphic minerals.

The significant features of stylolites are described below.

1 Stylolite architecture is shown in Fig. 7.8. Many waveforms are possible and the amplitude of the waveform is a measure of the amount of dissolution across the stylolite surface.

Fig. 7.7 Tectonic stylolite in limestone showing the development of a square waveform.

2 Stylolites generally form normal to σ_1, but oblique forms may have the relationship shown in Fig. 7.8f. Stylolites oblique to σ_1 may form in zones of layer-parallel slip and occur together with vein accretion steps. These are termed slickolites and may be used to determine the direction of movement on the slip plane (Figs. 7.8g and 5.6).

3 Stylolitization is favoured by small grain sizes, as in micritic limestones.

4 Stylolites parallel to bedding commonly have very irregular surfaces which will tend to prevent bedding plane slip from occurring, i.e. they lock the bedding planes together.

123

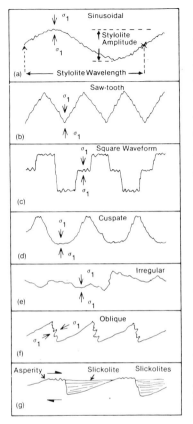

Description and measurement of stylolite plane orientations and the orientation of the stylolite columns will give valuable information on the microstructural history of the outcrop.

7.5 Mapping joints, veins and stylolites

In areas of complex structure, joints are generally not represented on the geological map, but they may be measured and recorded in the field notebook for subsequent analysis. In folded rocks, joints are best analysed in areas of uniform dip, i.e. on fold limbs (Fig. 7.4), and where the geometric relationships to the fold hinge can be observed directly. As well as the orientation, type of joint, and age relationships, you need to measure the dip of bedding, fold axes and fold axial plane orientations, so that the reference frame of a, b and c axes can be defined (Fig. 7.4).

Data that need to be collected for joint, vein or stylolite analysis are summarised in Table 7.1.

Fig. 7.8 Examples of stylolite architecture, showing its relationship to the σ_1 direction. (a) sinusoidal waveform showing wavelength and amplitude; (b) saw-tooth waveform; (c) square waveform; (d) cuspate waveform; (e) irregular waveform; (f) oblique saw-tooth waveform with σ_1 inclined to the stylolite plane; (g) slickolites showing the shearing along the stylolite plane to remove the asperities (bumps) by pressure solution and deposition of vein material in the hollows (see Fig. 5.6c).

Table 7.1 Data to be collected from observations on joints, veins and stylolites.

Structure	What to measure	What observations to record	Results of analysis
Joints J1 Extension Joints	Dip direction (or strike and dip) (Figs. 2.6 & 2.5).	Fracture type (dilational, shear, or hybrid). (Fig. 7.1).	For conjugate shear-fractures — stress systems.
Conjugate Shear Joints (Angle 60–90°)	Dip direction of conjugate fracture array (if developed). Line of intersection of conjugate arrays (Figs. 2.11–2.13).	Conjugate fracture system. (Fig. 7.2). Bedding and uniformity of bedding dip. Fracture spacing. Bed thicknesses. Length of fractures relative to bed thicknesses. Nature of fracture surface. (Fig. 7.3). Nature of fracture infilling (quartz; carbonate; fibrous or massive).	History of fracture movement. Bed competencies. History of fracture movement.
Conjugate Hybrid Joints (Angle <60°)	Line of bedding intersection on fracture plane (Figs. 2.11–2.13).		Gives apparent dip of bedding. Used to calculate true bedding attitude.
Additional information required for analysis.	Dip direction of bedding (fractures are best analysed in areas of uniform bedding). Orientation of fold axis. Orientation of fold axial plane.	Relationship of fracture to bedding. (Figs. 7.4, 7.5) Relationships of fractures to fold. Cylindrical (Fig. 7.4) Non-cylindrical (Fig. 7.5)	Analysis of fracture systems with respect to bedding and fold limbs. (Figs. 7.4, 7.5) Gives fracture systems: a–c, b–c, etc.

8
Polyphase deformation

If an area has undergone only one deformation, producing cylindrical folds, then the poles to bedding are generally distributed in a great circle girdle (Fig. 3.2); minor fold axes show constant orientation parallel to major fold axes (Fig. 3.8); the bedding/cleavage intersection is constant and parallel to the major fold axis (Fig. 4.3); and cleavage is relatively constant (it may fan about the fold axial plane) and approximates to the axial plane of the fold (Fig. 4.3). Such an area is *structurally homogeneous*.

For non-cylindrical folds (periclinal or conical folds produced by a single deformation, Figs. 3.6 & 7) we will find only parts of the map area that are structurally homogeneous. The axial planes of the folds (hence cleavages) will be relatively constant but the fold axes will show

systematic changes in their plunges (Fig. 3.6).

If the area has undergone more than one deformation, then the distribution of structural elements becomes more complex and the products of polyphase folding may be seen.

Polyphase folding is indicated by:

1. A wide distribution of bedding attitudes (away from simple great circle or conical patterns on the stereographic projection);
2. Fold interference patterns;
3. Folding of planar and linear structures that have been produced by earlier deformation;
4. Superposition of later fabrics on earlier fabrics (cleavages or schistosities).

Table 8.1 lists the terminology that is used in polyphase terranes.

Table 8.1 Terminology used in polyphase terranes. (Note: Bedding is denoted S_0

Deformation		Fold phase	Axial-planar foliation	Lineations (intersection of S_n/S_{n-1})
First deformation	D_1	F_1	S_1	$L_1 (S_1/S_0)$.
Second deformation	D_2	F_2	S_2	$L_2 (S_2/S_1)$. $L_0^2(S_2/S_0)$.
Third deformation	D_3	F_3	S_3	$L_3 (S_3/S_2)$. $L_0^3(S_3/S_0)$. $L_1^3(S_3/S_1)$.

8.1 Fold interference patterns

Polyphase deformation is recognised by the interference patterns produced in outcrops. Ramsay (1967) has recognised three basic end-members of a continuous series of fold interference patterns (Fig. 8.1) for two phases of folding, F_1 and F_2.

Type 1—'Egg box' or 'dome and basin' pattern. This pattern arises when both F_1 and F_2 fold axes and axial planes are orthogonal or at high angles to each other. *The F_1 axial planes remain unfolded* (Figs. 8.1a and 8.2a).

Type 2—'Mushroom' pattern. In this pattern some of the fold closures are attached to 'stalks', unlike the completely closed forms of Type 1. This occurs when the F_1 and F_2 fold axes and axial planes are *not* at a high angle to each other and the F_1 and F_2 axes are *not* coaxial. *The F_1 fold axial planes are folded. (Fig. 8.1b, 8.2b)*

Type 3—Refolded fold pattern *where the folding is coaxial* but the F_1 and F_2 axial planes are at a high angle. F_1 fold axial planes are folded. *(Fig. 8.1c, 8.2c)*

In many cases fold interference patterns are not obvious but are only revealed by detailed outcrop mapping.

8.2 Lineations in polyphase terranes

Polyphase deformation is characterised by a wide variation in the orientation of linear structures.

For two phases of folding, F_1 and F_2, the following are found:

8.2.1 F_2 fold axes

The orientation of the F_2 fold axes will depend upon the orientation of the F_1 fold limbs.

1 If the F_1 folds were *isoclinal* then the F_2 fold axes would be relatively constant.

2 If the F_1 folds were *not* isoclinal than F_2 fold axes would vary according to the orientation of F_1 fold limbs. In many cases the F_2 folds are of smaller wavelength than the earlier F_1 folds, and here the F_2 axes will define domains of constant F_2 orientation corresponding to particular limbs of the F_1 fold, and hence enable the location of the F_1 fold hinge lines to be established (Fig. 8.3).

Although you should expect significant variations in F_2 fold axis orientations, *the F_2 axial planes will be relatively constant in their orientation.*

8.2.2 Deformation by F_2 similar folding

In F_2 similar folding the movement may be thought of as taking place in the 'a_2' movement direction in the 'a_2-b_2' planes which are parallel to the axial plane of the F_2 fold. Thus, when a planar surface containing an L_1 lineation is affected by an F_2 similar fold (Fig. 8.4), the lineation L_1 remains in a plane, the orientation of which is controlled by the 'a_2' direction and the original orientation of the lineation L_1. The angle between the

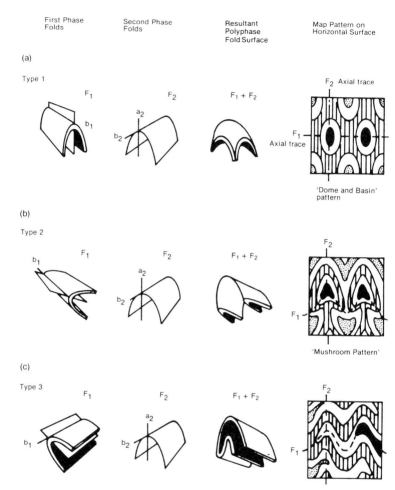

First Phase Folds Second Phase Folds Resultant Polyphase Fold Surface Map Pattern on Horizontal Surface

(a)

Type 1

F_1 F_2 $F_1 + F_2$ F_2 Axial trace

b_1 a_2 F_1 Axial trace

b_2

'Dome and Basin' pattern

(b)

Type 2

F_1 F_2 $F_1 + F_2$ F_2

b_1 a_2 F_1

b_2

'Mushroom Pattern'

(c)

Type 3

F_1 F_2 $F_1 + F_2$ F_2

b_1 b_2 F_1

Refolded Folds

Fig. 8.1 Fold interference patterns: (*a*) Type 1 patterns are formed by superposition of two orthogonal fold sets and produce a dome and basin pattern on a horizontal map surface; (*b*) Type 2 patterns produced by the superposition of a *non-coaxial* second upright fold set upon an inclined first-phase fold pattern. Mushroom style patterns are produced on a horizontal map surface; (*c*) Type 3 patterns produced by *coaxial* refolding of an inclined first-phase fold by an upright second-phase fold. Refolded folds are produced on a horizontal map surface.

128

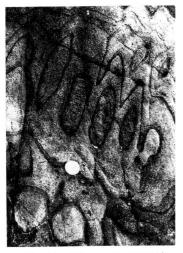

Fig. 8.2a Type 1 'dome and basin' fold interference pattern in polydeformed psammites.

Fig. 8.2b Type 2 'mushroom style' interference pattern in polydeformed psammites. Field of view *ca.* 50 cm.

Fig. 8.2c Type 3 refolded fold interference patterns in polydeformed psammites.

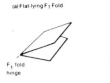

(a) Flat-lying F_1 Fold

F_1 fold hinge

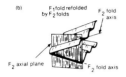

(b)

F_1 fold refolded by F_2 folds

F_2 fold axis

F_2 axial plane

F_2 fold axis

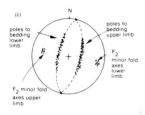

(c)

N

poles to bedding lower limb

poles to bedding upper limb

F_2 minor fold axes lower limb.

F_2 minor fold axes upper limb

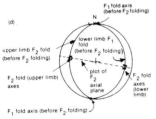

(d)

F_1 fold axis (before F_2 folding)

N

lower limb F_1 fold (before F_2 folding)

upper limb F_2 fold (before F_2 folding)

F_2 fold (upper limb) axes

plot of F_2 axial plane

F_2 fold axes (lower limb)

F_1 fold axis (before F_2 folding)

Fig. 8.3 The orientation of the second-phase fold axes (F_2) controlled by the orientation of the first-phase fold limbs. The stereographic projection shows the plot of the second-phase fold axes F_2 (crosses) as a result of the intersection of the second-phase fold (F_2) axial plane and the first-phase fold limbs. Note that the poles to bedding (π plots) define two great circles — the poles to which correspond to the locations of minor fold axes.

deformed lineation and the similar fold axis varies systematically over the fold hinge (Fig. 8.4). Because the deformed lineation remains in a *plane*, it will be redistributed about a great circle in the stereographic projection (Fig. 8.4). This can be recognised *in the field* by placing your map board on the deformed lineation and attempting to line up the other parts of the lineation within the plane of the board—if they can be aligned then the lineations lie in a plane, and hence the redistribution was generated by similar style folding.

8.2.3 Deformation by F_2 flexural slip folding

In this case the orientation of the lineation L_1 changes as the layers slide over one another (folding is parallel but the lineation retains a constant angle to the F_2 fold axis, see Fig. 8.5). Hence the lineation will adopt a curved form and plot as small circles on the stereographic projection (Fig. 8.5). (Note that the two cases described above are end-members of a spectrum of fold mechanisms, see Ramsay (1967) for a full discussion.)

8.3 Sub-areas

In a mapping area where polyphase deformation exists, the subdivision into *structurally homogeneous sub-areas* is *essential* in order to analyse the overall structure. Mapping should always be carried out with this in

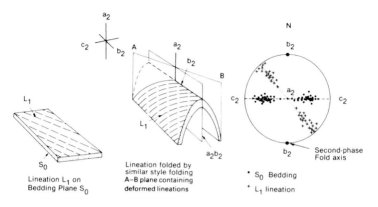

Fig. 8.4 F_2 similar style folding, redistributing an earlier L_1 intersection lineation. L_1 remains in a plane that contains the second-phase movement axis a_2. The stereographic projection shows the distribution of S_0 (solid circles) by F_2 and the redistribution of L_1 (crosses) in a great circle. The tectonic axes a_2, b_2 and c_2 for the second deformation are also shown.

mind, and you should attempt to divide your area into structurally homogeneous sub-areas *whilst in the field.* Only then can you seek out key exposures to confirm or disprove your interpretation.

Analysis of structural data using the stereographic projection alone is inadequate in areas of polyphase deformation because this method does not take into account the geographic location of structures.

Division of a mapping area into structurally homogeneous sub-areas is based upon defining the following:

1 Areas where there is a constant orientation of a particular generation of lineation (Fig. 8.6);

2 Areas where there is a constant orientation of a particular foliation plane;

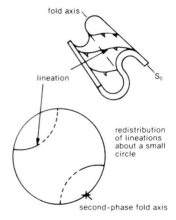

Fig. 8.5 F_2 flexural slip folding, redistributing an L_1 lineation which produces a small circle pattern in the stereographic projection.

131

3 Areas where there is a constant fold axis orientation, F_1 or F_2 etc. (Fig. 8.6);

4 Areas distinguished by the axial surfaces of the various fold structures, i.e. vergence boundaries.

Sub-areas can be determined by inspection from a map, i.e. recognition of interference patterns and changes in lineation orientations (Fig. 8.6). They can be analysed by careful plotting of data on the stereographic projection, taking great care to distinguish the *geographic location* of the data on the stereographic projection.

8.4 Mapping polyphase terranes

When mapping in terranes that show evidence of polyphase deformation the following procedure should be adopted.

1 Identify and describe the fold styles and collect data as described in chapter 3 and Table 3.4. In particular examine outcrops for evidence of interference patterns (Figs. 8.1 and 8.2) and overprinting relationships; examine fold hinges to determine cylindricity (Fig. 3.7) and to see if lineations are folded around the fold hinges. The style of folding and of interference relationships will usually be reflected in the outcrop patterns on your map and in your cross-section (see 9.2).

2 Observe the cleavages present in the outcrops and collect the appropriate data (Tables 4.1 and 4.2). In particular look for superposition of cleavages and foliages and record age relationships (younger cleavages cut older cleavages). Map and record the orientations of the cleavages/foliations throughout the map area.

3 Observe, map and record the structural relationships (Tables 5.1–5.3) of the lineations developed in the map area. In particular distinguish between intersection lineations parallel to fold axes and mineral stretching lineations often at high angles to the fold axes. Identify lineations associated with particular fold phases.

4 Map out lithological boundaries to determine whether interference patterns (Figs. 8.1 and 8.2) are developed.

5 Identify sub-areas in the field. In particular the orientation of intersection lineations (parallel to fold axes) is most often used to define sub-areas (Fig. 8.6) hence you must systematically map and record lineation data in your map area.

Table 8.2 summarises the structural data to be collected in polyphase terranes when two major phases of folding are present. If more than two deformation phases can be recognised then structural data for the subsequent foliations (S_3, S_4 etc.), lineations (L_3, L_4 etc.) and minor folds must also be collected.

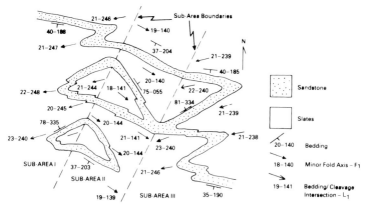

Fig. 8.6 Map of sandstones and slates showing polyphase folding. Sub-areas are identified using uniformity of the plunge of F_1 minor fold axes and the L_1 bedding/cleavage intersection lineations (generated by F_1 folding). Note that the sub-area boundaries define the traces of the F_2 fold axial surfaces.

Table 8.2 Summary of data to be collected when mapping in polyphase terranes (two major phases of folding).

Structure	What to Measure	What Observations to Record	Results of Analysis
F_1 structures			
S_0 bedding.	Dip direction.	Nature of S_0; sedimentary structures. Way up.	Complex pattern; redistribution by F_2.
S_1 First foliation.	Dip direction.	Nature of S_1, vergence, facing. Relationship to S_2 and overprinting relationships.	Folded in Type 2 and 3 interference patterns. (Fig. 8.1)
L_1 (S_1/S_0).	Plunge.	Nature of L_1.	Folded by F_2. (Figs. 8.4, 8.5)
ML_1 mineral stretching lineation.	Plunge.	Nature of ML_1.	Folded by F_2. Figs. 8.4, 8.5
F_1 fold axis.	Plunge.		Folded by F_2. (Figs. 8.4, 8.5)
F_1AP (F_1 axial plane).	Dip direction.		Folded in Type 2 and 3 (Figs. 8.1, 8.2) interference patterns.
F_2 structures			
S_2 second foliation.	Dip direction.	Nature of S_2, overprinting relationships.	Constant orientation. (Fig. 8.3)
L_2 (S_2/S_1).	Plunge.	Nature of L_2, overprinting of L_1?	Approximately constant orientation (Fig. 8.3).
L_0^2 (S_2/S_0).	Plunge.	Nature of L_0^2.	Parallel F_2— variable.
ML_2 mineral stretching lineation.	Plunge.	Nature of ML_2.	
F_2 fold axis.	Plunge.		Variable depending upon first-phase fold limbs. (Fig. 8.3)
F_2AP (F_2 axial plane).	Dip direction.		Constant. (Fig. 8.3)

134

9

First steps in overall interpretation and analysis

This chapter is a brief introduction to the interpretation and analysis of field maps, of field structural observations and orientation data. The construction of cross-sections and report writing are reviewed. Throughout this Handbook emphasis has been placed upon:

(1) accurate and detailed *description* of structures;
(2) accurate *measurement* of structures;
(3) *Identification of structural style* — often reflected in the minor structures (Table 1.1, sections 3.9 and 8.4).

During your mapping programme, whilst still in the field, you must always carry out an *ongoing interpretation of your map and structural data*, and *construct draft (preliminary) cross-sections*. These will enable you to identify key problem areas and collect additional information necessary for a more precise interpretation.

9.1 Map patterns and map interpretation

Your geological field map and field notebook contain the essential data upon which the structural interpretation is based. When you have mapped a significant part of your area you should begin your interpretation and analysis using the following procedure (note that the interpretation should be continuously updated and modified whilst you are in the field) —

1 In the field, plot onto your topographic base map all of the outcrop localities and the appropriate structural data. This may involve plotting from aerial photograph overlays using the correction techniques outlined by Barnes (1981). This field map should be an *outcrop map* that also shows boundaries observed in the field (Fig. 9.1a). This and the notebook are critical as they allow you to assess the geological data base — how much exposure, where, and what structural data have been measured.
Plot on your field map the axial surface traces of major folds, the plunges of major folds (derived from stereographic analysis — Fig. 3.2) and plot the *major faults* (Fig. 9.1a).

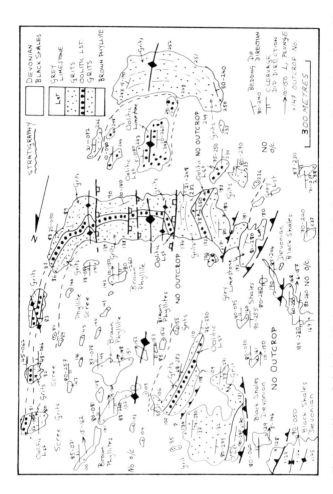

Fig. 9.1a Field map of a faulted anticlinal structure showing the distribution of outcrops and boundaries as observed in the field. (Note — topographic contours and some structural data have been omitted for clarity in reproduction).

2   Construct a *contact or boundary map* (Fig. 9.2b) by extrapolating observed boundaries and interpreting your field map. This should take into account the structural style of the area (sections 3.9, 8.4 and Table 1.1).

3   If a cleavage or cleavages are developed you should construct a *cleavage or S surface map* (in this example, Fig. 9.1c — the cleavage map has been combined with a lineation map) that shows cleavage orientation e.g. S_1, S_2. This kind of map is extremely useful in polyphase terranes as it will indicate refolding of earlier cleavages, rotation by shear zones and orientations of the fold axial surfaces.

4   Construct a *lineation map* (in Fig. 9.1c cleavage and lineation maps are combined) which shows the distribution of linear structures — *minor fold axes, intersection lineations and mineral stretching lineations*. This is essential in areas of polyphase folding in order to establish structurally homogeneous *sub-areas* (see section 8.3 and Fig. 8.6).

5   Construct a *summary structural map* (Fig. 9.1d) on which the major faults and folds are extracted from the field map and plotted on a transparent overlay. This often aids interpretation and allows the fault and fold systems to be analysed (see chapters 3 and 6).

6   In areas where more than one phase of deformation has been identified, *sub-areas* should be identified and their boundaries plotted on the contact of boundary map. Analysis of orientation data from structurally homogeneous sub-areas (section 8.3) will allow major fold plunges to be plotted on the map.

7   Construct preliminary cross-section(s) (e.g. Fig. 9.1e) using the techniques outlined in section 9.2 below.
    Map interpretation and cross-section construction are interactive processes that must be carried out together.

Note: (a) Topographic effects on the outcrop pattern must always be taken into account. For example Fig. 9.2 shows the effects of topography on the location of fold axial traces, particularly for plunging folds where the axial trace is located upslope away from the position of maximum curvature in the outcrop pattern (Fig. 9.2f).

(b) The idealized fold interference patterns shown in Fig. 8.1 are for both sets of folds having the same *wavelength* and *amplitude*. This is often *not* the case and commonly the second set of folds will be smaller in wavelength and amplitude than the first set. Thus the interference patterns will be modified from the textbook case as shown in Fig. 8.1.

(c) In areas where direct measurement of the orientations of large planar features such as faults, unconformities and planar contacts are not possible, construction of structure contours will enable the strike and dip to be determined (Fig. 9.3) and hence provide

137

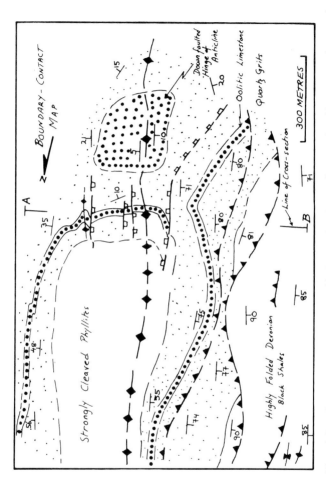

Fig. 9.1b Boundary or contact map constructed by interpreting Fig. 9.1a above. Stratigraphic contacts and structural features have been extrapolated between outcrops. Legend and symbols are the same as in Fig. 9.1a.

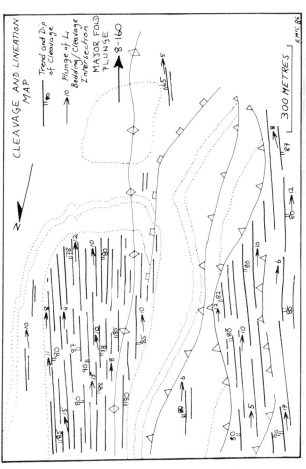

Fig. 9.1c Cleavage and lineation map constructed from Fig. 9.1a. Lines indicating the trend and dip of the cleavage are shown together with arrows indicating the plunges of the bedding/cleavage intersection lineation L_1. Note that both the cleavage and lineations are almost exclusively developed in the phyllites and the shale units (Figs 9.1a and 9.1b).

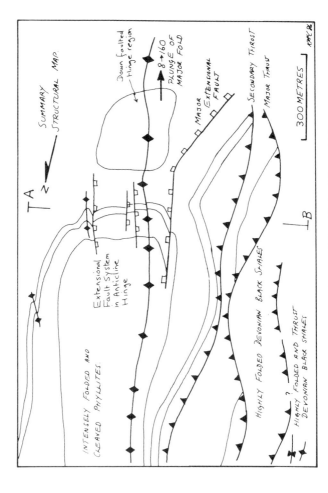

Fig. 9.1d Summary structural map showing the major folds and faults of the area in Fig. 9.1a.

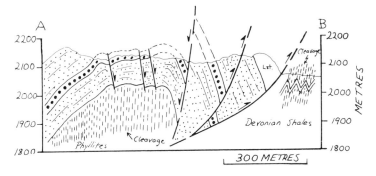

Fig. 9.1e Preliminary cross-section along A-B (Fig. 9.1b). Cleavages and fault movements are shown. Legend is the same as Fig. 9.1a.

valuable data for the construction of cross-sections.

9.2 Cross-sections

Cross-sections are an essential part of a structural synthesis. During mapping sketch cross-sections (approximately scaled) should be drawn in your field notebook (Fig. 9.4), and preliminary, correctly scaled cross-sections along traverses, etc. should be made as part of an on-going interpretation. *All cross-sections* should show the form lines of all structural features S_0, S_1, S_2 etc (e.g. Fig. 9.1e).

The cross-sections representing the final interpretation for your report should be drawn with *as much care*, and have *as much detail* as your map. They should be constructed using the following procedure —

1 Determine the structural grain in your area — is there a dominant set of folds or a dominant set of faults — e.g. one major direction of thrusting? If folds are the dominant structures determine the plunges of the major folds (stereographic analysis — Fig. 3.2). If thrust or normal faulting is the major feature determine the direction of slip — i.e. the movement direction (see section 6.2). Always carefully locate your line(s) of section so that they are — (a) perpendicular to the plunge of the major folds; or (b) parallel to the slip direction of the major faults (ie. parallel to the direction of tectonic transport).

In some polyphase terranes where there is more than one direction of major folding, it may be appropriate to construct additional cross-sections perpendicular to the axes of the

141

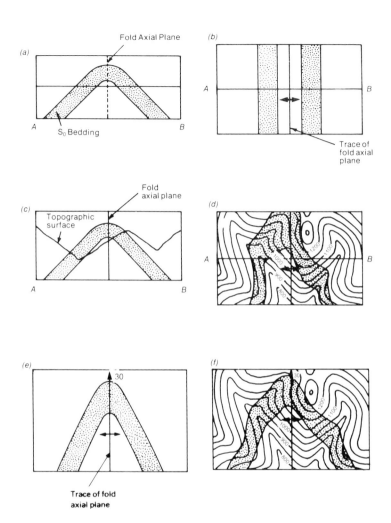

Fig. 9.2 Outcrop patterns of folds, showing the effects of topography on the map patterns: (a) horizontal upright fold in the profile plane; (b) horizontal upright fold exposed on a horizontal surface; (c) horizontal upright fold in cross-section in an area of topographic relief; (d) horizontal upright fold exposed in an area of topographic relief; (e) plunging upright fold exposed on a horizontal plane surface; (f) plunging upright fold exposed in an area of topographic relief. Note the displacement of the fold axial trace upslope from the surface closure on the map. (Adapted from Ragan, 1985 and reproduced with permission.)

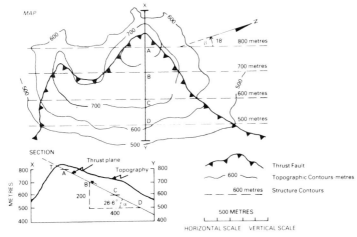

Fig. 9.3 Construction of structure contours and determination of the dip and strike of a planar structure. The map shows the outcrop of a thrust plane in an area of topographic relief. *Structure contours* for the thrust plane are constructed by joining points where the thrust plane intersects a particular topographic contour e.g. 700 metres. The 700 metre structure contour (dashed line) defines the strike of the thrust plane at 700 metres elevation in this area. On the cross-section X-Y the topographic surface is plotted and the outcrop of the thrust plane T is marked. The structure contours for the thrust plane intersect the line of cross section (X-Y) in the points A, B, C, and D and these are plotted on the cross-section to define the thrust plane. The strike of the thrust plane is given by the angle between north and the strike of the structure contours $\beta = 18°$, and the dip can be calculated from the cross-section — angle of dip = $\tan 0.5 = 26.6°$. Strike of the thrust plane is $018°$ north east and dip is $26.6°$ (i.e. $27°$) south east.

other fold sets. In areas of complex extensional or thrust faulting it may be appropriate to draw cross-sections parallel to the strike of the major structures — i.e. *perpendicular to the tectonic transport direction* in order to highlight along strike variations.

2 Always draw your cross-section(s) through areas on your map that contain good structural and lithological data (i.e. not through large patches of unexposed geology). Data off the line of section should be projected down plunge (see 9.2.2 and Fig. 9.6) into the plane of the cross-section with corrections for apparent dip (Appendix I) and for thickness changes (Appendix II).

3 *Always* draw cross-sections with the horizontal and vertical scales *equal*. Where possible they should be at the same scale as the map but in some cases they may need to be enlarged to illustrate complex structures.

4 Plot all S surfaces (S_0, S_1, S_2 etc.) on the cross-section. Mark on the movement directions of all faults.

143

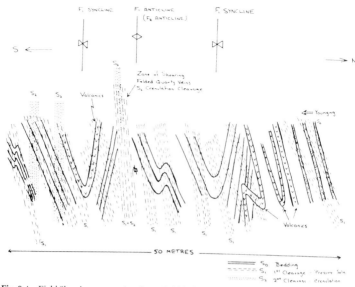

Fig. 9.4 Field Sketch cross-section through folded pelites and psammites. Note the form lines for bedding (S₀), and cleavages (S₁ and S₂). Horizontal and vertical scales are approximately equal.

5   Plot younging, vergence and facing directions on cross-sections.

6   Draw your cross-sections to reflect the *structural style* of the area (see Table 1.1, sections 3.9 and 8.4). The structural style is often reflected in the minor or mesoscale structures in outcrop.

7   Where possible sections should be balanced (Dahlstrom, 1969) — see section 9.2.3. Section balancing is not always possible but should always be considered.

9.2.1 *Vertical cross-sections*

Vertical cross-sections are the most commonly constructed cross-

section. In areas where the folds are non-plunging these structure sections give profile views of the folds, *however*, in areas of plunging folds vertical structure sections give distorted views of the folds. In this case down plunge projections are required to give inclined structure sections normal to the fold axis — i.e. in the profile plane of the folds.

For an area with non-plunging folds (Fig. 9.5), the line of cross-section is positioned perpendicular to the trend of the fold axes where there is relatively abundant orientation information and fold structures of interest. Once the line of section has been chosen, A–A' on Figure 9.5, a topographic profile is constructed at the same scale as the map using the elevation control from contour lines. The formation contacts and their dips are marked on the profile and orientation data (strike and dip) *are projected* onto the line of section, using apparent dips where necessary (Appendix 1). The cross-section is then constructed taking into account the observed structural style in the field — e.g. parallel folding (constant bed thickness) or similar style folding. Form lines of other S surfaces (S$_1$, S$_2$ etc.) as well as bedding and formation contacts should be plotted on the section (Fig. 9.4).

9.2.2   Inclined structure sections

Where the folds are plunging, cross-sections should be constructed

normal to the fold axis — i.e. in the profile plane of the fold. These are *inclined structure sections*. The following steps should be taken for constructing inclined structure sections in which the orientation and boundary data are projected down the plunge of the fold:

(1) Determine the trend and plunge of the fold axis usually using a π pole diagram.

(2) Draw the line of section on the map at 90° to the plunge of the fold axis i.e. line X–Y. (Fig. 9.6). The section plane will be orientated at 90° to the fold axis.

(3) If there is little or no topography then individual points can be projected onto the plane of

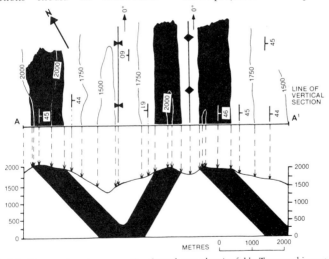

Fig. 9.5 Construction of a cross-section through *non-plunging* folds. Topographic contours are projected onto the line of section to construct the topographic profile. The bedding dip angles are then projected onto the topographic profile, and the section is constructed (in this case assuming a near chevron fold style and constant bed thickness). Horizontal scale equals vertical scale.

the cross-section — using the formula:

$di = dm (\sin \alpha)$ (Fig. 9.6c), where di is the distance down the inclined plane, dm = map distance and α = plunge of the fold axis. This projection is carried out for each point e.g. point P to P′ on the map until a down plunge projection of the fold is constructed (Fig. 9.6d). (4) If there is appreciable topography (Fig. 9.6e) then structure contours must be drawn for the line of section on the map and also on the inclined structure section (Fig. 9.6e). Then the above formula (Fig. 9.6c) can be applied using the distance dm as the

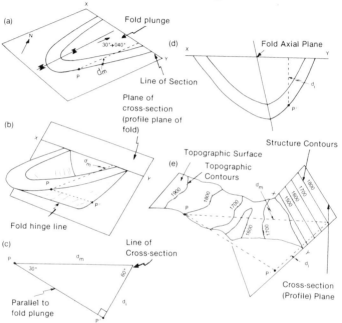

Fig. 9.6 Construction of a down-plunge projection through a plunging fold (i.e. 30° to 040°). (a) Map view of the plunging fold on a horizontal surface. The line of cross-section X-Y is drawn at 90° to the fold plunge. (b) 3D view of the plunging fold showing the profile plane (containing the section line X-Y) and showing the projection of point P on the map to point P′ on the cross-section (profile) plane. d_m is the distance (parallel to the fold plunge) on the map plane between point P and the line of cross-section. (c) Geometric contruction to calculate the distance d_i, by which P must be projected down the cross-section plane to point P′. (d) Resultant down-plunge projection of the fold in the plane of cross-section that is 90° to the fold hinge line. (e) 3D view of the projection of point P from a surface with topographic relief. Structure contours of the cross-section plane are drawn on both the map and the projection plane. Point P is projected to P′ by calculating the distance, d_m, from P at 1800 m on the map to the profile plane at 1800 m, and then using the calculation in (c).

146

distance between the map point at a particular elevation and the structure contour on the inclined structure section at the same elevation (Fig. 9.6e). The distance di is then projected down the section plane and the cross-section constructed as before.

Once a down plunge projection has been constructed the true profile section is revealed. The cross-section can be balanced or dip isogons can be constructed.

9.2.3 Balanced cross-sections

Balanced cross-sections are generally constructed for deformed sedimentary sequences where the stratigraphy is known and well developed — for example foreland fold and thrust belts. In high grade terranes and polyphase deformation with penetrative foliations the construction of balanced sections is difficult if not impossible.

Basic assumptions for balancing cross-sections

(1) Sections are constructed perpendicular to the fold axes — i.e. down-plunge projections.
(2) Sections are constructed parallel to the tectonic transport direction.
(3) We assume no *volume loss* (or gain) during deformation.
(4) We assume that there is little elongation or contraction in the strike direction — perpendicular

to the tectonic transport direction — i.e. no movement in or out of the plane of the cross-section.
As a corollary we assume that there is a state of 2D plane strain in the section parallel to the tectonic transport direction.
(5) The area of the deformed section is the same as the undeformed area — i.e. no area change.

The nett result is a geometrically feasible cross-section that is restorable. The restored section shows the original stratigraphy and the fault trajectories.

There are two basic methods of drawing balanced cross-sections: (a) assuming *constant line length*, and (b) assuming *constant area*. In areas where *parallel folding* occurs e.g. flexural slip folding — constant bed thickness then we use *line length balancing*. In areas where there are *similar* style folds and *cleaved* rocks we use *area balancing*.

The simplest case of section balancing, *line length balancing,* can be attempted in camp. To measure constant line lengths we can use either a piece of string or a curvimeter (Fig. 9.7) — we assume constant bed thickness.

In order to construct a balanced section we need to:

(1) Assemble the field data on the topographic profile for the cross-section — projecting data in the plane of the cross-section where required (Fig. 9.8a) (the line of section should be normal to the

Fig. 9.7 Curvimeter for measuring line lengths on maps and cross-sections.

fold plunge of the major structures);

(2) Determine the stratigraphic thicknesses in the area and construct a stratigraphic template (Fig. 9.8b).

(3) Establish a *pin line* — where there is no interbed slip and the beds are pinned together. *The pin line* is usually located in the foreland (Fig. 9.8b), where there is no deformation, or on the axial surface of the anticline.

(4) *Measure away from the pin line* producing a restored section at the same time as constructing the balanced section. The restored section should have *no gaps or overlaps* and the fault trajectories should be reasonable (Fig. 9.8c). For staircase fault geometries the *ramp angles* of faults should be less than or equal to 30°. The cross-section should end on a pin line as shown in Fig. 9.8c.

9.3 Report writing

An essential element of the structural mapping programme is writing a report that communicates the results of your mapping and analysis in a clear, unambiguous and concise fashion. Before you leave the field ensure that you have, in addition to descriptions and interpretations of the stratigraphy, sedimentology, metamorphism and igneous rocks, the following:

1 Descriptions of the major structural features — folds and fault patterns.

2 Descriptions of each of the structural elements and of their geographic distributions — e.g. S_0, S_1, S_2 etc., L_1, L_2 etc. and minor structures — folds and faults; in some cases additional maps (e.g. Fig. 9.1) may be required.

3 The relative ages of structures based upon superposition and interference relationships.

4 Justification for the division of the map area into sub-areas of structural homogeneity (Section 8.3).

5 Relationships between deformation and metamorphism; in particular descriptions of metamorphic minerals and their relationships to tectonic fabrics and of fault rocks (see Fry, 1984).

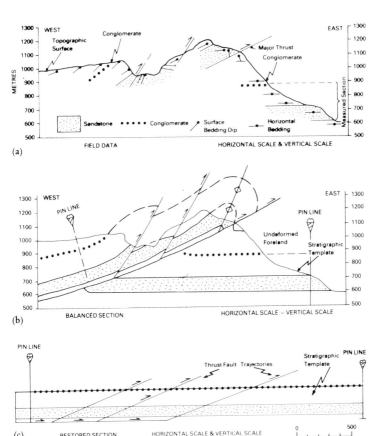

Fig. 9.8 Construction of a balanced cross-section through a simple fold-thrust structure.
(a) Field data is plotted on the topographic profile (using down plunge projections (Fig. 9.6) and apparent dips (Appendix 1) where required). Bedding dips, lithological and tectonic boundaries are shown together with the location of a measured section used to determine stratigraphic thicknesses.
(b) Balanced section constructed at the same time as the restored section — Fig. 9.8c. The stratigraphic template — Fig. 9.8c, was based on the measured section (Fig. 9.8a). The section has been balanced using *line length balancing* and the assumptions outlined in the text (section 9.2.3). The *front pin line* has been placed in the undeformed foreland and the *end pin line* has been placed in a convenient segment of stratigraphy at the western end of the section. Faults have been projected to depth using their surface dips.
(c) Restored section on the stratigraphic template. The thrust fault trajectories and the front and end pin lines are shown.

149

6   Relationships to regional structures and a synthesis of the structural evolution of the area.

7   Stress and strain analysis based upon measurements of fracture patterns and deformed objects.

8   Kinematic analysis — tectonic transport directions and structural evolution (e.g. dominant movement patterns of thrust faults etc.).

Your report should be fully illustrated with maps, stratigraphic columns, cross-sections, fully annotated diagrams and photographs.

Good luck!

References and further reading

References to take to the field

BARNES, J. W., (1981) *Basic Geological Mapping*. Geological Society of London Handbook Series, 1. Open University Press, 112 pp.

BOYER, S., and ELLIOTT, D., (1982) 'Thrust systems'. *Bulletin of the American Association of Petroleum Geologists*, 66, 1196–1230.

DAVIS, G. H., (1984) *Structural Geology of Rocks and Regions*. New York, Wiley, 492 pp.

FRY, N., (1984) *The Field Description of Metamorphic Rocks*. Geological Society of London Handbook Series, 3. Open University Press, 110 pp.

HANCOCK, P. L., (1985) 'Brittle microtectonics: principles and practice'. *Journal of Structural Geology*, 7, 437–458.

PHILLIPS, F. C., (1971) *The Use of Stereographic Projection in Structural Geology*. 3rd ed. London, Edward Arnold, 90 pp.

RAGAN, D. M., (1985) *Structural Geology: an Introduction to Geometric Techniques*. 3rd ed., New York, Wiley, 393 pp.

THORPE, R. S., & BROWN, G. C., (1985) *The Field Description of Igneous Rocks*. Geological Society of London Handbook Series, 4. Open University Press, 162 pp.

TUCKER, M. E., (1982) *The Field Description of Sedimentary Rocks*. Geological Society of London Handbook Series, 2. Open University Press, 124 pp.

Further reading

ANDERSON, E. M., (1951) *The Dynamics of Faulting*. Edinburgh, Oliver and Boyd, 241 pp.

BADGELY, P. C., (1959) *Structural Methods for the Exploration Geologist*. New York, Harper, 280 pp.

BELL, A. M., (1981) 'Vergence: an evaluation'. *Journal of Structural Geology*, 3, 197–202.

BUTLER, R. W. H. (1982) 'The terminology of structures in thrust belts'. *Journal of Structural Geology*, 4, 239–45.

DAHLSTROM, C. D. A., (1969) 'Balanced cross sections'. *Canadian Journal of Earth Sciences*, 6, 743–57.

HOBBS, B. E., MEANS, W. D., & WILLIAMS, P. F., (1976) *An Outline of Structural Geology*. New York, Wiley, 571 pp.

HUDDLESTON, P. J., (1973) 'Fold morphology and some geometrical implications of theories of fold development'. *Tectonophysics*, 16, 1–46.

PARK, R. G., (1983) *Foundations of Structural Geology*. Glasgow, Blackie, 135 pp.

PRICE, N. J., (1966) *Fault and Joint Development in Brittle and Semi-Brittle Rocks*. Oxford, Pergamon, 176 pp.

RAMSAY, J. G., (1967) *Folding and Fracturing of Rocks*. New York, McGraw-Hill, 567 pp.

RAMSAY, J. G., (1974) 'Development of chevron folds'. *Geological Society of America, Bulletin*, 85, 1741–54.

RAMSAY, J. G., (1980) 'Shear zone geometry: a review'. *Journal of Structural Geology*, 2, 83–99.

RAMSAY, J. G., (1982) 'Rock ductility and its influence on the development of tectonic structures in mountain belts', in HSU, K. (ed.) *Mountain Building Processes*. London, Academic Press, 111–128.

RAMSAY, J. G. & GRAHAM, R. H., (1970) 'Strain variation in shear belts'. *Canadian Journal of Earth Sciences*, 7, 786–813.

RAMSAY, J. G. and HUBER, M. I., (1983) The Techniques of Modern Structural Geology. Volume 1: Strain Analysis. London, Academic Press, 307 pp.

SIBSON, R. H., (1977) 'Fault rocks and fault mechanisms'. *Journal of the Geological Society of London*, 133, 191–214.

SIMPSON, C. & SCHMID, S. M., (1983) 'An evaluation of criteria to deduce the sense of movement in sheared rocks'. *Geological Society of America, Bulletin*, 94, 1281–1288.

WILSON, G. (with COSGROVE, J.) (1982) *An Introduction to Small-Scale Geological Structures*. Allen and Unwin, 128 pp.

WILLIAMS, G. D., & CHAPMAN, T. J., (1979) 'The geometrical classification of non-cylindrical folds'. *Journal of Structural Geology*, 1, 181–186.

Appendix I

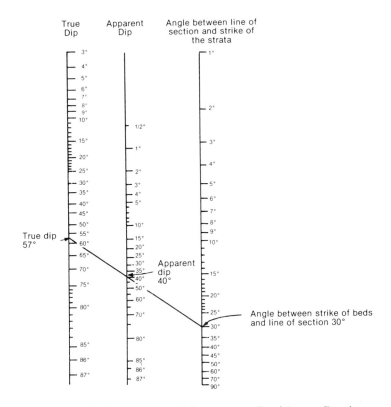

True Dip

Apparent Dip

Angle between line of
section and strike of
the strata

True dip
57°

Apparent
dip
40°

Angle between strike of beds
and line of section 30°

Fig. I.1 A nomogram for determining the true dip from an apparent dip and vice versa. Example: on a cross-section the beds have an apparent dip of 40° and the line of the cross-section is at 30° to the strike of the beds (as seen on the map)—on the nomogram mark off the difference in strike (30°) and the angle of apparent dip (40°), and then draw a straight line through these points to give the true dip—57°.

Appendix II

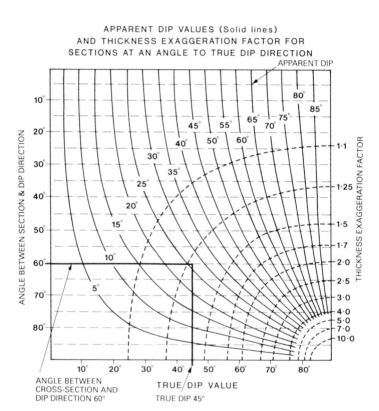

APPARENT DIP VALUES (Solid lines)
AND THICKNESS EXAGGERATION FACTOR FOR
SECTIONS AT AN ANGLE TO TRUE DIP DIRECTION

Fig. II.1 A graph to determine the apparent dip angles (solid curved lines) and the thickness exaggerations (dashed curved lines) for bedding in cross-sections which are constructed at an angle to the true dip direction (i.e. not at 90° to the strike). Example: for bedding with a true dip of 45° and for a cross-section oriented at 60° to the dip direction; from the graph the apparent dip is 27° and the thickness of bedding will be 1.275 × the true thickness.

154.

Appendix III
Strain measurements

Guidelines for identifying and measuring the state of strain in outcrops or hand specimens are briefly outlined. Finite strain determinations in rocks may be achieved by measuring the distortions in a variety of markers for which the pre-deformation geometry is known, e.g. ooliths, fossils and pebbles.

The main types of strain markers are:

1 Initially *spherical* objects — e.g. ooliths, accretionary lapilli, reduction spots.
2 Initially *ellipsoidal* objects — e.g. pebbles or sedimentary grains.
3 Known initial angles between lines — e.g. fossils with known angular relationships, trilobites, graptolites, brachiopods etc.
4 Known initial lengths or ratios of lengths of lines, e.g. belemnites, crinoid stems, boudin structures.

Strain determinations may be made using:

(*a*) direct measurement in the field — i.e. axial ratios of deformed pebbles.

(*b*) photographing deformed objects and then analysing the photographs in the laboratory.

(*c*) collecting oriented samples and then analysing these in the laboratory.

Comprehensive accounts of the laboratory techniques for finite strain determination in rocks are given in Ramsay and Huber (1983) and Ragan (1985). Any geologist wishing to carry out detailed strain analyses must be familiar with these texts before commencing fieldwork.

Initially spherical or elliptical strain markers

If you can identify the principal planes of finite strain (XY, XZ, YZ planes — Fig. III.1) in the field, then direct measurement of deformed objects in these planes will allow you to estimate the finite strain. For example, it is generally accepted that, as a first approximation, the foliation or slaty cleavage is parallel to the XY plane of the finite strain ellipsoid, then measurement of the axial ratios of deformed reduction spots (Fig. III.2a) in the cleavage plane and on a joint plane normal to the cleavage (Fig. III.2b) will allow the axial ratios of the finite strain ellipsoid to be calculated in the field (Fig. III.2c).

Similarly you may determine the deformation of pebbles in a con-

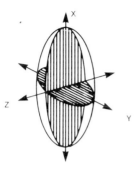

Fig. III.1 The strain ellipsoid. The principal strain directions are X— the long axis, Y— the intermediate axis and Z the short axis. The planes containing these axes are called principal planes — i.e. the cleavage plane approximates to the XY plane of the finite strain ellipsoid.

glomerate by measuring (in the field) their axial ratios on the cleavage/foliation plane, and on joints as close as possible to other principal planes of finite strain. Deformed pebbles, ooliths etc. may not have been initially spherical, therefore you need to photograph or sample the exposure in order to make accurate strain determinations in the laboratory.

Initially cylindrical strain markers

In most cases cylindrical markers cannot be successfully used for strain

Fig. III.2a Natural strain ellipse of a deformed reduction spot on the cleavage plane of a slate (i.e. the XY plane of the finite strain ellipsoid).

Fig. III.2b Deformed reduction spots on a joint plane in slate. The joint plane is 90° to the slaty cleavage plane and parallel to the long axis of the finite strain ellipsoid: i.e. it is the XZ plane of the finite strain ellipsoid.

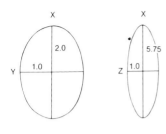

FLINN DIAGRAM

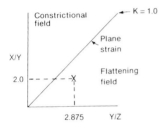

Fig. III.2c Determination of the finite strain ellipsoid from measurements on the deformed reduction spots in Figs. III.2a & b. In Fig. III.2a the ratio X/Y is 2·0:1, whereas in Fig. III.2b the ratio X/Z has an average of 5·75:1. Therefore the ratio Y/Z can be calculated as 5·75/2·0 = 2·875:1, — therefore X : Y : Z = 2·0 : 1·0 : 0·348. These ratios are plotted on a Flinn diagram of X/Y versus Y/Z which shows the field of flattening and the field of constriction (assuming no volume changes in the rock). The straight line between these two fields indicates a condition of plane strain where X = 1/Z and there is no change in the length of the Y strain axis.

measurements except in the case of 'Skolithos' or 'Monocraterion' worm burrows in quartzites. In undeformed sediments these are circular in plan on the bedding plane and the long axis of the burrow is usually 90° to the bedding. Upon deformation, the worm tubes become elliptical in the bedding plane — thus allowing measurement of layer-parallel shortening in the bedding (Fig. III.3a). This *axial ratio* can be measured directly in the field, together with the orientation of the long axis of the strain ellipse and the *axes* of the worm tubes which have become sheared in the bedding (Fig. III.3b) thus allowing the determination of *angular shearing strain*.

Known angular relationships

For single fossil forms the angular shear strain can be calculated from deformed angular relationships (Fig. III.4) e.g. deformed brachiopods, bivalves, graptolites etc. For lines initially at 90° and now at an angle φ the angular shear strain γ is calculated as $\gamma = \tan(90 - \varphi)$.

Known initial length of lines

Boudin structures or deformed fossils such as belemnites (Fig. III.5) can be easily measured to calculate the amount of extension in the plane of the structure. The strain ratio is simply *(final length–initial length)/(initial length)*. Measurements can be made in the field or samples collected for laboratory analysis. Note however, that the strain measured is likely to be the minimum finite strain, because structures such as boudinage occur where a competent unit is deformed in a more highly deformed, ductile or less competent matrix.

For precise strain determinations field data should be augmented by laboratory analysis of either (a) photographs of deformed markers, (b) *collection of oriented samples* (Section 2.7), or (c) field measurements analysed further in the laboratory.

Table III.1 summarises the commonly applied techniques used to determine finite strains in rocks.

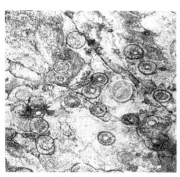

Fig. III.3a Bedding plane in deformed quartzites, showing what were initially circular bedding plane sections of *Monocraterion* trace fossils, now deformed into strain ellipses with a strain ratio in the bedding plane of 1·3:1.

Fig. III.4 Deformed lamellibranch with a small angular shearing strain.

Fig. III.3b Section perpendicular to bedding in deformed quartzites showing sheared *Skolithos* trace fossils — initially perpendicular to the bedding surface and now sheared to an angle of 45° to the bedding — angular shear strain of $\gamma = 1\cdot0$.

Fig. III.5 Boudinaged belemnite on the bedding plane. The elongation or stretch of the belemnite can be determined by measuring the stretched length L_1 and summing the lengths of all of the individual segments of the belemnite to give the initial length L_0. The elongation is then $(L_1 - L_0)/L_0$)

Table III.1 Data to be collected for finite strain determinations.

STRAIN MARKER	MEASUREMENTS	METHODS OF ANALYSIS AND RESULTS (See Ramsay and Huber 1983)
A INITIALLY SPHERICAL OBJECTS Ooliths, pisoliths, accretionary lapillae, concretions, reduction spots, spherical fossils.	METHOD 1. DIRECT FIELD MEASUREMENT 1.1 *Orientation* of the principal planes of the finite strain ellipsoid–cleavage plane = XY plane, plus 2 orthogonal joint sets if possible. Measure orientations of the axes of the ellipses in these planes. (Figs. III.1, III.2) 1.2 Measurement of axial ratios of deformed ellipses on each principal plane. 1.3 For deformed objects lying in the bedding plane measure the orientation of bedding, bedding/cleavage intersection and the ellipse ratios in the bedding plane. Trace of cleavage defines the XY plane of the finite strain ellipsoid. METHOD 2. PHOTOGRAPHING THE DEFORMED OBJECTS 2.1 Measure orientations as in 1.1. 2.2 *Mark* and measure orientation of a reference line in each plane to be photographed — then photograph. Be sure to record all data in your field notebook.	Combine data from the XY, YZ and XZ planes to obtain the final ellipsoid ratio X:Y:Z. Note: often it will only be possible to measure the strain accurately in one principal plane, e.g. the cleavage-XY plane. Plot strain ellipses on map and cross-section. Analyse photograph in laboratory, using centre-centre, Fry method or Rf/φ technique (objects not usually perfectly spherical). Combine strain data from the three principal planes, as in METHOD 1.

Table III.1—(Cont'd on next page)

159

Table III.1—(Cont'd)

STRAIN MARKER	MEASUREMENTS	METHODS OF ANALYSIS AND RESULTS (See Ramsay and Huber 1983)
	METHOD 3. COLLECTION OF ORIENTED SAMPLE	
	3.1 Record orientation of bedding, cleavage and bedding/cleavage intersection.	Reorient sample in laboratory. Identify principal strain ellipsoid planes.
	3.2 Mark and record reference orientation as in Section 2.7.	Slab along XY, YZ, and XZ planes. Analyse using techniques in METHOD 2.
	3.3 Collect oriented sample.	
B INITIALLY ELLIPTICAL OBJECTS Conglomerate pebbles and sedimentary grains.	DIRECT FIELD MEASUREMENT	Plot Rf/φ diagram —determine finite strain ellipse for each principal plane.
	4.1 Measure orientations as in 1.1.	
	4.2 As in METHOD 1, above.	Combine data from principal planes.
	PHOTOGRAPHING DEFORMED PEBBLES	
	As in METHOD 2, above.	Analyse photograph in laboratory using Fry or Rf/φ techniques.
		Combine data from the three principal planes, as in METHOD 1.
	COLLECTION OF ORIENTED SAMPLES	
	As in METHOD 3, above.	As in Method 3. Analyse using Fry or R φ methods.
		Combine data from the three principal planes, as in METHOD 1.

160